Volume 2

RAMAN SPECTROSCOPY

Theory and Practice

Volume 2

RAMAN SPECTROSCOPY
Theory and Practice

Edited by Herman A. Szymanski

Dean of the College
Alliance College
Cambridge Springs, Pennsylvania

Springer Science+Business Media, LLC 1970

Library of Congress Catalog Card Number 64-23241

ISBN 978-1-4684-3029-5   ISBN 978-1-4684-3027-1 (eBook)
DOI10.1007/978-1-4684-3027-1

© 1970 Springer Science+Business Media New York
Originally published by Plenum Press, New York in 1970
Softcover reprint of the hardcover 1st edition 1970

Preface

Raman Spectroscopy, Volume 1, was conceived to provide integrated and comprehensive coverage of all aspects of the field by a group of specialists. However, in the three years since the first volume was published much important work has been done. Since *Volume 1* was very well received, this second volume has been prepared in the belief that an extension of the coverage it offers will satisfy a real need in this rapidly changing and extremely interesting field.

Any pretension to comprehensive coverage, however, had to be abandoned. In order to keep the material in a work of this nature up to date, a cutoff date has to be set. Inevitably one or two of the planned articles fail to materialize by this deadline, and other interesting topics may come into focus too late to permit the preparation of a worthwhile discussion by the target date. Still, in fairness to those authors who kept to the schedule, the cutoff date has to be enforced, even though this means sacrificing breadth of coverage to timeliness.

I wish to thank all the contributors to this volume for their effort, their cooperation, and their punctuality, and it is my hope that the policy I have followed will result in the presentation of current thought on a series of interesting aspects of the subject of Raman spectroscopy.

May 1970   H.A.S.

Contents

Chapter 1

Vibrational Rules of Selection and Polarization: Their Practical Uses and Limitations

L. A. Woodward

University of Oxford
Oxford, England

INTRODUCTION

The vibrational selection rules for the Raman effect and for infrared absorption and the rule of polarization for vibrational Raman lines are based solely on symmetry considerations. For a molecule whose structure (and hence whose symmetry point group) is known, application of the rules gives information as to the number of fundamental vibrational frequencies permitted in either type of spectrum and the number whose Raman lines will be polarized.

If we are concerned with a molecule of unknown structure for which two (or more) models with different symmetries can be reasonably proposed, we can make predictions of the above kinds for each. In general, the predictions for the rival models will be different, so that by experimental observation of the actual spectra it will be possible, in principle, to discriminate between the proposed types of structure and to decide which is the right one for the molecule in question.

It is important to note that this method, being based solely on general symmetry theory, is quite independent of the special nature of the molecular force field. For this reason it is, in principle, a method of great power; for a complete description of the force field is generally inaccessible. Indeed, the method has often been used with notable success.

Nevertheless, it is true that in certain cases circumstances may arise which limit its usefulness. It is the object of this chapter to discuss the method in general, and to illustrate both its power and its limitations by examples from the literature.

1

VIBRATIONAL SELECTION RULES

Before discussing applications, it will be convenient to give a brief résumé of the derivation of the selection rules. We shall concern ourselves throughout only with fundamentals, because generally speaking they are most easily observable. Corresponding selection rules can be derived for overtones and combination tones, but these usually appear with relatively low intensity, especially in Raman spectra.

The intensity of a transition is proportional to the square of the relevant transition moment, and so the condition that a transition be allowed is that the transition moment shall not vanish.

Infrared Absorption

For a fundamental transition (vibrational quantum number change from 0 to 1) the transition moment for infrared absorption is given by

$$\mu_{01} = \int \psi_0 \mu \psi_1 \, dQ$$

where ψ_0, ψ_1 are, respectively, the wave functions of the initial and final states, μ is the electric dipole moment of the molecule as a function of Q, and Q is the normal coordinate of the vibrational mode. The integral is to be extended over the whole coordinate range.

We must bear in mind that the dipole moment, being a vector, has three components (μ_x, μ_y, μ_z in a Cartesian system), and that, in greater detail, an integral of the above kind applies separately to each.

The condition that the integral for the component μ_i (where i denotes either x, y, or z) shall not vanish is that the integrand $\psi_0 \mu_i \psi_1$ shall be totally symmetric, i.e., shall be transformed into itself by all the symmetry operations of the molecular point group. Since ψ_0 (ground vibrational state) is known always to be totally symmetric, it follows that the product $\mu_i \psi_1$ must be totally symmetric. This is only the case if both factors belong to the same symmetry species. For the transition to be permitted, it will suffice if this is so for at least one of the components of μ. Now it is also known that ψ_1 always belongs to the same symmetry species as does the vibration itself. We can therefore state the infrared-absorption selection rule for fundamentals as follows:

A fundamental is permitted in infrared absorption only if its species is the same as that of at least one of the components of the electric dipole moment.

The species of the dipole-moment components are the same as those of the corresponding translations, and these are customarily given in the point-group character tables. It is therefore a simple matter to read off from these tables the selection rules for vibrations of any species.

Raman Effect

In considering Raman scattering we are not concerned at all with the intrinsic dipole moment μ of the molecule, but only with the dipole moment P which is *induced* in the molecule by the electric field E of the incident light. This is given by

$$P = \alpha E$$

where α is the molecular polarizability.

For a fundamental Raman transition the transition moment is, accordingly,

$$P_{01} = \int \psi_0 \alpha E \psi_1 \, dQ = E \int \psi_0 \alpha \psi_1 \, dQ$$

Here we must remember that α is a *tensor* quantity representable by an array of nine components (α_{xx}, α_{xy}, etc.). However, because the tensor is a symmetric one (which means that $\alpha_{ij} = \alpha_{ji}$), only six of the nine components are distinct. In greater detail, a transition-moment integral of the above form applies separately to each component.

Just as in the case of infrared absorption (see above), we can state the selection rule for Raman scattering as follows:

A fundamental is permitted in Raman scattering only if its species is the same as that of at least one of the components of the polarizability.

A component α_{ij} transforms in the same way as does the product of the translations T_i and T_j, and the species of the components of α (or in certain cases suitable linear combinations of them) are customarily given in the point-group character tables. It is therefore an easy matter to read off the selection rules for vibrations of any species.

POLARIZATION RULE FOR RAMAN SCATTERING

For a fluid sample, when the incident light is natural (i.e., un-polarized) and the scattering is observed at right angles to the incident direction, the degree of depolarization ρ_n of a Raman fundamental is given by

$$\rho_n = \frac{6(\gamma_{01})^2}{45(\bar{\alpha}_{01})^2 + 7(\gamma_{01})^2}$$

where $\bar{\alpha}_{01}$ and γ_{01} are, respectively, the mean value and anisotropy invariants of the transition tensor α_{01}. It follows that the line is only polarized (i.e., has $\rho_n < \frac{6}{7}$) if $\bar{\alpha}_{01}$ does not vanish.

Now

$$\bar{\alpha}_{01} = \tfrac{1}{3}[(\alpha_{xx})_{01} + (\alpha_{yy})_{01} + (\alpha_{zz})_{01}]$$

$$= \int \psi_0 \bar{\alpha} \psi_1 \, dQ$$

where $\bar{\alpha}$ is the corresponding mean value invariant of the molecular polarizability tensor α. It is known that $\bar{\alpha}$, being of the same symmetry as $T_x^2 + T_y^2 + T_z^2$, always belongs to the totally symmetric species. It follows that $\bar{\alpha}_{01}$ can be different from zero only if ψ_1 is also totally symmetric, i.e., if the vibration concerned belongs to this species. We can therefore state the Raman polarization rule for fundamentals as follows:

Only fundamentals of totally symmetric vibrations give polarized Raman lines ($\rho_n < \frac{6}{7}$). The Raman lines of fundamentals belonging to all other species are depolarized, i.e., have $\rho_n = \frac{6}{7}$.

PREDICTIONS FROM THE RULES

A nonlinear molecule containing N nuclei has $3N - 6$ normal vibrational modes. For a linear molecule the corresponding number is $3N - 5$.

For any proposed symmetry (i.e., point group) the symmetry species of the modes can be deduced. Where degenerate species are involved, pairs of modes (or sets of three) will be such that their members will necessarily have identically the same frequency. In general, therefore, the number of *distinct* fundamental frequency values will be less than the total number of normal modes. In any case the number of distinct frequencies can be deduced.

By using the rules of selection and polarization it is possible, for any proposed point group, to make the following predictions:

1. The number of fundamental frequencies which will be permitted in the infrared absorption spectrum.

2. The number of fundamental frequencies which will be permitted in the Raman spectrum.
3. The number of coincidences, i.e., the number which will be permitted in both types of spectrum.
4. The number of fundamentals in the Raman spectrum which will be polarized.

For different proposed models the sets of predictions will in general be different. This is the basis of the experimental method of determining which model is the right one in any particular case.

ROUGH ESTIMATES OF FUNDAMENTAL FREQUENCIES

As already pointed out, the power of the above method depends on the fact that its application does not require any knowledge of the force field. In using the method, we are concerned with the *numbers* of permitted fundamentals and the *number* of polarized Raman lines, and not at all with their frequency values. Indeed, frequency measurements are, in principle, not required.

Nevertheless, it is of some advantage, in practice, to have in mind, in a qualitative way, the approximate ranges in which the different fundamental frequencies may be reasonably expected to lie. Thus, the conclusion that the numbers of observed spectroscopic features favor a certain molecular model may receive an extra measure of support from the observation that these features do in fact lie at roughly the anticipated positions on a frequency scale.

Due caution is required, however, since the estimates of frequencies must necessarily be only approximate. The actual values are determined partly by the nuclear masses involved and partly by the nature of the force field; and the latter is not precisely known. We may be able to set up a more or less plausible approximation to it, realizing from experience that bond-stretching force constants are, in general, considerably larger than angle-change force constants, and perhaps by invoking values that have been found for similar bonds and bond angles in related molecules. Insofar as the field so constructed represents a fair approximation of the actual field, the values of the fundamental frequencies calculated from it will give a fair idea of the actual values. In the nature of things, however, we have no reliable way of showing whether the field we have used is in fact a good approximation to the truth.

For any proposed point group it is possible to derive the symmetry coordinates for each set of symmetrically equivalent internal-displacement coordinates (such as stretchings of a particular type of bond or changes of a particular type of bond angle). The actual normal coordinates of a chosen symmetry species will, of course, be "mixtures" (i.e., linear combinations) of all the symmetry coordinates of that species; but if the reduced masses or the force constants involved are expected to be considerably different for the different symmetry coordinates, then the degree of mixing will be expected to be correspondingly small. In these circumstances it will be a fair approximation to regard each normal mode as involving only one symmetry coordinate, and the sort of frequency value to be expected for the latter will give a fair idea of the actual frequency. On the other hand, where the possibility of considerable mixing is foreseen, the situation will be less clear and the forecast approximate frequency values will be less informative. Nevertheless, we may hope that for each fundamental that is permitted (or forbidden) we shall know roughly the nature of the mode and the frequency range within which it should appear in (or be absent from) the vibrational spectra.

Because of the relative magnitudes of force constants, fundamentals of modes involving mainly bond stretching will, in general, tend to lie at higher frequencies than those involving mainly angle-change deformations. Of course, the nuclear masses will play an important part. Thus, a deformation frequency for the change of the angle between two bonds to very light atoms may well be higher than a stretching frequency where the bond in question joins two relatively heavy atoms.

One type of molecule is of special interest in this connection, and some examples will be discussed later in this chapter. It is the type in which a number of hydrogen atoms (for example those in methyl groups) are attached to a "skeleton" consisting wholly of relatively heavy atoms. Here the symmetry coordinates concerned with motions of the very light hydrogen nuclei will mix very little with those concerned with motions of the heavy skeletal nuclei. It will, therefore, be a good approximation to think of the normal modes as being either vibrations of the methyl groups (considered as distinct entities) or vibrations of the skeleton (considered separately from the hydrogens). In general, frequencies of the methyl-group vibrations will lie considerably higher than those of the skeletal vibrations, so that there will be little difficulty in distinguishing between the two in the actual spectra. If any doubts on this score should arise, they can be effectively dispelled by studying the effect upon the frequencies of replacing all the hydrogen atoms by deuterium.

In seeking to determine the symmetry of a molecule of this type, one will be concerned mainly with the symmetry of the skeleton. It will be possible to obtain evidence as to this by considering the skeletal fundamentals only and by applying to them the rules of selection and polarization, just as if the hydrogens were not present or (otherwise expressed) just as if each methyl group were a single atom.

SIMPLE EXAMPLES OF STRUCTURE DETERMINATION

This section will outline some examples of the application of the vibrational spectroscopic method for the determination of molecular symmetry. For two reasons, only very simple examples will be chosen : first, in order to illustrate as clearly as possible the nature of the method and, second, because we shall have occasion to refer back to these simple cases in a later discussion of more complicated cases which are formally analogous.

The Mercurous Ion in Aqueous Solution

This example must surely represent the utmost in simplicity. The question to be decided is whether the ion exists in the simple form Hg^+ or (as suggested by some physicochemical evidence) in the double form Hg_2^{2+}.

If monatomic, the ion can have no vibration, whereas if diatomic it will have just one totally symmetric stretching mode. Here the infrared spectrum can give no information, since the single fundamental of the homonuclear diatomic model (dipole moment, zero) is forbidden. In the Raman effect, on the contrary, it is permitted as a polarized line. Moreover, Raman spectroscopy (unlike infrared) is applicable without difficulty to species in aqueous solution.

Because of the large mass of the Hg nuclei, the frequency of Hg_2^{2+} will be expected to be quite low. Assuming a stretching force constant of about the usual magnitude for a single bond, the value is likely to be roughly in the region of 200 cm^{-1}.

In fact, as long ago as 1934, the writer found[1] that an aqueous solution of mercurous nitrate showed (in addition to features attributable to the solvent and the NO_3^- ion) a single intense, polarized Raman line at $\Delta v = 169$ cm^{-1}. This provides convincing proof that the mercurous ion has the structure $(Hg-Hg)^{2+}$. Incidentally, this was the first example of the vibration of a metal–metal bond, a type that has subsequently attracted a considerable amount of interest.

In the same work the possibility that the thallous ion might likewise be diatomic was investigated. Certain physicochemical observations had been interpreted on the basis of the structure Tl_2^{2+}. However, no Raman frequency attributable to such a double ion could be observed. In view of the positive success with the mercurous ion, this negative result may be taken as strong evidence that the thallous ion has the simple structure Tl^+.

Molecules of Type XY_2

After diatomic species the next simplest are triatomic XY_2, in which both Y atoms are attached to the central X atom. Here we have two possibilities with different symmetries: the linear model (point group $D_{\infty h}$) and the bent model (point group C_{2v}). For each of these the number of distinct fundamental frequencies is three.

The predictions from the rules of selection and polarization are given in Table 1.

Table 1. Predictions for Fundamentals of Symmetrical XY_2 Species

Model	Number permitted in infrared absorption	Number permitted in the Raman effect	Number of coincidences	Number of polarized Raman lines
Linear $(D_{\infty h})$ Y–X–Y	2	1	0	1
Bent (C_{2v}) X / \ Y Y	3	3	3	2

For the linear model the number of coincidences between the infrared and Raman spectra is zero. This is an example of the so-called *alternative rule* or rule of mutual exclusion—a special aspect of the selection rules applying only to molecules which (like linear Y–X–Y) possess a center of symmetry. It may be stated as follows:

For any molecule with a center of symmetry, fundamentals that are permitted in infrared absorption are forbidden in the Raman effect, and fundamentals that are permitted in the Raman effect are forbidden in the infrared. It should, however, be noted that in certain cases some fundamentals may be forbidden in both kinds of spectrum.

For both models (linear and bent) two of the fundamentals involve bond stretching (mainly or entirely), and the third involves angle-change deformation. As to expected frequencies, the two stretches (symmetric and asymmetric) will lie in the same relatively higher range, and the deformation in a lower range.

Generally speaking, the predictions of Table 1 are well realized in practice for molecular species whose structures are known from other evidence. As an example of a linear symmetric XY_2 molecule we may cite *mercuric chloride*, $HgCl_2$. The Raman spectrum of the vapor[2] consists of a single line ($\Delta v = 355\,cm^{-1}$). A single line (polarized) of similar frequency is observed for the compound in the molten state and in solution in various solvents. This fundamental (symmetric stretching) is not observed in the infrared absorption spectrum of the vapor[3] which, however, shows a fundamental at $413\,cm^{-1}$ (asymmetric stretching). The bending frequency is also permitted in the infrared. It was not observed for the vapor in the work quoted, because its low frequency lay beyond the range of the spectrometer used. From a study of the infrared spectrum of the solid compound[4] it appears that the frequency in question is in fact $74\,cm^{-1}$.

As an example of a bent symmetrical XY_2 molecule we may cite *sulfur dioxide*, SO_2. Here all three fundamentals are observed in the infrared spectrum[5] of the gas (frequencies approximately 520, 1150, and $1300\,cm^{-1}$). All three are also found in the Raman spectrum of the liquid[6]; the two lower (bending and symmetric stretching) are polarized and the highest (asymmetric stretching) is depolarized. Thus, the predictions of Table 1 are well borne out.

The vibrational spectroscopic method has been of special value in the question of the structure of *xenon difluoride*.[7] Only one Raman line is observed for the vapor ($\Delta v = 515\,cm^{-1}$). This frequency is absent from the infrared absorption spectrum, which shows fundamentals at 213 and $557\,cm^{-1}$. This evidence strongly indicates a linear structure for this very interesting molecule.

Another case where the vibrational evidence has been decisive is that of *nitrous oxide*, N_2O. This molecule is isoelectronic with CO_2, which is known to be symmetrical and linear. It was therefore reasonable to suppose that N_2O had the same sort of structure, linear NON. Indeed this view appeared to receive support both from the fact that the molecular dipole moment is practically zero, and also from the electron-diffraction evidence that the molecule is undoubtedly linear and that, as far as could be determined, the two bond lengths are practically equal. However, the infrared spectrum[8] shows three strong bands which are undoubtedly fundamentals (589, 1285, and 2224 cm^{-1}). Moreover, the two higher of these frequencies also appear in the Raman spectrum of the gas and both are polarized.[9] The coincidences in the two types of spectrum rule out at once the possibility that the molecule has a center of symmetry. As it is known to be linear, the only remaining structure is the nonsymmetrical NNO. (Owing to the similar electron-scattering properties of the two different kinds of atom, a distinction between this structure and the symmetrical NON would be difficult by the electron-diffraction method.)

The structure NNO belongs to the point group $C_{\infty v}$, for which the predictions from the rules of selection and polarization are given in Table 2.

Table 2. Predictions for Fundamentals of Species Y–Y–X

Model	Number permitted in infrared absorption	Number permitted in the Raman effect	Number of coincidences	Number of polarized Raman lines
Linear $(C_{\infty v})$ Y–Y–X	3	3	3	2

The experimental findings are consistent with the predictions. It is true that only two fundamentals are actually observed in the Raman effect, whereas all three are permitted. The missing one is the deformation. This is not so surprising in the light of general experience that deformation fundamentals not infrequently have lower intrinsic Raman intensities than do stretching fundamentals. Also, very weak features are more likely to be unobservable with gaseous samples than with liquids, where the molecular population density is so much higher. In the case of N_2O the deformation frequency must be of such low intensity as to escape observation, for there is no conceivable structure which would permit only the two stretching fundamentals as polarized lines while at the same time forbidding the deformation.

Molecules of Type XY_3

Further simple examples may be found among molecules of type XY_3 in which all three Y atoms are attached to the central X atom. Here the simplest possible models are the planar (point group D_{3h}) and the pyramidal (point group C_{3v}). For both the number of distinct fundamental frequencies is four. The predictions are given in Table 3.

Table 3. Predictions for Fundamentals of Symmetrical XY_3 Species

Model	Number permitted in infrared absorption	Number permitted in the Raman effect	Number of coincidences	Number of polarized Raman lines
Planar (D_{3h})	3	3	2	1
Pyramidal (C_{3v})	4	4	4	2

For the planar model, where the rules are the more restrictive, it is specially notable that the totally symmetric stretching fundamental, which is permitted in the Raman effect as the sole polarized line and expected to have a high Raman intensity, is the one which is forbidden in the infrared.

Here again the predictions from the rules of selection and polarization are well realized in practice for molecules whose structures are known from other evidence. As a planar example we may take *boron trifluoride*, BF_3. The infrared absorption spectrum[10] shows three fundamentals (482, 720, and 1505 cm^{-1}). In the Raman spectrum only two are observed,[11] whereas three are permitted (Table 3); but it is important to note that the more intense of the two observed Raman lines ($\Delta v = 888$ cm^{-1}, polarized) is absent from the infrared. This is clearly the totally symmetric stretching frequency. The other observed Raman feature agrees with the infrared absorption at 482 cm^{-1} (degenerate deformation). The failure to observe 1505 cm^{-1} (degenerate stretching) in the Raman effect must be due to the relatively very low intensity.

With the boron isotopes present in natural abundances (^{10}B:^{11}B about 1:4), the infrared permitted fundamentals are all double. The values quoted above are for ^{11}BF$_3$. We shall return later to this isotopic effect and the decisive evidence it gives in this case on the question of the symmetry of the molecule.

An example of a pyramidal XY$_3$ molecule is *phosphorus trichloride*, PCl$_3$. All four fundamentals are observed in the Raman spectrum of the vapor and the liquid[12] (Δv for vapor, 184, 256, 482, and 514 cm^{-1}). All four also appear in the infrared spectrum.[13]

Of special interest among molecules of type XY$_3$ is *chlorine trifluoride*, ClF$_3$. In contrast to the examples cited above, this does not conform with the predictions for either the symmetrical planar or the pyramidal model. The infrared spectrum of the gas[14] shows six fundamentals (326, 364, 434, 528, 703, and 752 cm^{-1}). Two of these are observed also in the Raman spectrum of the gas,[14] and four of them in the Raman spectrum of the liquid. Clearly the molecule must have a lower symmetry than either D_{3h} or C_{3v}. In fact, the vibrational findings (although incomplete for the Raman spectrum) are consistent with the very unusual T-shape indicated by microwave investigation.[15] In this, two of the fluorines are stereochemically equivalent and the third is different. The point group is C_{2v}, for which there are six distinct fundamental frequencies, all permitted both in infrared absorption and in the Raman effect.

Molecules of Type XY$_4$

Here the simplest and commonest models are the regular tetrahedral (point group T_d) and the plane square (point group D_{4h}). For the former the number of distinct fundamentals is four, and for the latter, seven. The predictions from the rules of selection and polarization are given in Table 4.

For the regular tetrahedral model it is specially noteworthy that the sole totally symmetric stretching vibration (permitted in the Raman effect as a completely polarized line and expected to have a high Raman intensity) is forbidden in the infrared. The plane square model has a center of symmetry, which forbids any coincidences between the two types of spectrum. For this model one of the seven distinct fundamental frequencies is forbidden both in the infrared and the Raman effect.

Table 4. Predictions for Fundamentals of XY$_4$ Species

Model	Number permitted in infrared absorption	Number permitted in the Raman effect	Number of coincidences	Number of polarized Raman lines
Regular tetrahedral (T_d) Y — X — Y Y Y	2	4	2	1
Plane square (D_{4h}) Y Y — X — Y Y	3	3	0	1

There are many examples of regular tetrahedral XY$_4$ species, including all the tetrahydrides and tetrahalides of Group IV elements and numerous tetra-halo complex ions with other central atoms. The isoelectronic pair *germanium tetrachloride*, GeCl$_4$, and the *tetrachlorogallate ion*, GaCl$_4^-$, are typical. The Raman spectra of both (GeCl$_4$ as liquid[16] and GaCl$_4^-$ in aqueous solution[17]) show the expected four fundamentals, the most intense being highly polarized. The observed frequencies have very similar patterns: for GeCl$_4$, 132, 172, 396 (polarized), and 453 cm^{-1}; for GaCl$_4^-$, 114, 149, 346 (polarized), and 386 cm^{-1}. We shall have occasion later to refer again to the findings for GaCl$_4^-$.

Very different patterns are found, for example, in the case of the plane square *tetrachloroplatinous ion*, PtCl$_4^{2-}$. The infrared spectrum of the solid potassium salt[18] shows three fundamentals (93, 183, and 320 cm^{-1}). The Raman spectrum of the ion in solution[19] also shows three (164, 304, and 335 cm^{-1}), the highest of which is the most intense and is polarized. The absence of coincidences is notable. The spectra clearly accord with the predictions for a plane square configuration.

Vibrational spectroscopy has been of great value in resolving the question of the molecular structure of *xenon tetrafluoride*, XeF$_4$. The infrared spectrum of the vapor[20] shows three strong fundamentals (123,

291, and 586 cm^{-1}). In the Raman spectrum of the vapor[17] two fundamentals have been observed ($\Delta v = 513$ and 549 cm^{-1}), neither of which coincides with an infrared band. This evidence excludes the regular tetrahedral model and provides strong support for a plane square structure for the free molecule, as indicated by x-ray diffraction studies of the solid compound.

In the case of the *sulfur tetrafluoride* molecule, SF_4, the vibrational spectroscopic evidence proves that the structure cannot be either of the two simple types so far discussed, in both of which all the four halogen atoms are symmetrically equivalent. The Raman spectrum of the liquid[21] shows at least seven lines, indicating a structure of relatively low symmetry. In the frequency range accessible to the infrared spectrometer used,[21] six bands were found which could reasonably be taken as fundamentals. Five of them represent coincidences with Raman features. The SF_4 molecule has a valency shell of 10 electrons, as compared with a group IV tetrafluoride which has only eight. It is apparent from the vibrational spectra that the "lone pair" in SF_4 is affecting the shape of the molecule. It is reasonable to suppose that the type of hybridization (including the lone pair) is essentially trigonal bipyramidal, in which case the lone pair might occupy either one of the three "equatorial" positions or one of the two "polar" positions. The resulting point groups would be, respectively, C_{2v} and C_{3v}. Predictions from the rules of selection and polarization are different for the two models, and in principle the vibrational spectroscopic method should be capable of deciding which model is correct. In practice, however, the experimental findings for this "difficult" compound were clearly incomplete, and a definite conclusion was not possible. A critical weighing of the evidence was judged to favor the C_{2v} model. Confirmation of the correctness of this view was subsequently forthcoming from studies by other methods.

The case of SF_4 illustrates that the full potential of the vibrational spectroscopic method may not always be realizable. Indeed, the impression given by the preceding simple examples may well have been unduly favorable. In fact, there are a number of circumstances which tend to detract from the practical usefulness of the method, and even among species of quite simple structures it is by no means an easy matter to select cases that are completely free from complications of one sort or another. In order to get a more realistic view, we will now enumerate and briefly discuss the principal sources of difficulty and doubt that may be encountered.

COMPLICATIONS AND LIMITATIONS

As already pointed out, the method is concerned with the numbers of permitted fundamentals and the numbers of polarized Raman lines. Any circumstances, therefore, that tend to give a wrong impression of these numbers will necessarily tend to falsify the deductions from them. In certain cases the numbers may appear to be greater than predicted for the correct model. Generally speaking, however, this is not so serious a matter as when, maybe owing to limitations of experimental techniques, the observed numbers are smaller than they should be.

Circumstances Tending to Make the Numbers Appear Too Large

The presence in the spectra of permitted *overtones* or *combination tones* may lead to error in deciding the numbers of permitted fundamentals. In Raman spectra, where overtones and combination tones are usually very weak and seldom observable, this is less likely than in the infrared, where they may be relatively stronger and so more likely to be mistaken for fundamentals. In what precedes we have confined our discussion to the rules of selection and polarization for fundamentals; but corresponding rules (likewise based solely on symmetry considerations) may be deduced for the overtones and combination tones of any molecule of known point group. Where there is doubt as to whether an observed feature is really a fundamental, it is often possible to throw light on the question by measuring its frequency (as well as those of the undoubted fundamentals), and then taking due account of the rules as they apply to the possible overtone or combination-tone assignments. It may not be possible, however, thus to dispel all doubt.

There is one special circumstance involving an overtone or combination tone which may cause the appearance (even in the Raman spectrum) of an "extra" frequency with an intensity which may be of the same order as that of a permitted fundamental. A notable example is found in the Raman spectrum of carbon dioxide. As is well known, this is a symmetrical linear molecule, for which the Raman selection rules permit only the totally symmetric stretching fundamental. Thus, the spectrum is expected to consist of a single intense, polarized line. The molecule is indeed such a simple one that complications would hardly be anticipated. Nevertheless, the spectrum is found[22] to consist of *two* strong, polarized lines of practically equal intensity ($\Delta v = 1285$ and $1388 \, \text{cm}^{-1}$). They cannot both be fundamentals, for the total number of these is only three, and (in accordance with the selection rules) the infrared absorption spectrum shows two (bending and asymmetric stretching) at 667 and 2349 cm^{-1}, respectively. It is clear that the symmetric stretching fundamental has appeared in the Raman spectrum as two lines instead of the expected one. This is a striking example of the phenomenon known as *Fermi resonance*. We note that the mean of the frequency values of the two Raman lines, 1336 cm^{-1}, is almost exactly twice the value of the bending fundamental, as found in the infrared. Now although the bending mode is forbidden in the Raman effect as a fundamental, its first overtone is permitted. Of course, but for Fermi resonance, it would have very low intensity and would probably be unobservable. In fact, however, its symmetry contains a component of the same totally symmetric species as the Raman-permitted stretching fundamental. In consequence of the very near equality of the frequencies, the two enter into Fermi resonance, in the sense that the wave functions of the vibrational states become mixtures of the individual wave functions. As a result the levels, which would otherwise be almost coincident, suffer mutual "repulsion," i.e., one goes higher and the other lower. Transitions from the ground state to both of them are permitted in the Raman effect, and hence the two observed lines. When, as in this case, the unperturbed frequencies are closely matched, the intensities of the resulting Raman doublet components are nearly equal. Loosely speaking, the displaced overtone is said to have "borrowed" intensity from the (displaced) fundamental. Where the frequency matching is less close, the extent of the repulsion of the levels is smaller and the difference of intensity between the components of the Fermi doublet is more marked.

Another familiar case of Fermi resonance is encountered in the Raman spectrum of carbon tetrachloride, where the highest fundamental (degenerate stretching) appears double. This is explained without difficulty as due to resonance with a combination of two lower fundamentals which belongs to the same symmetry species. Generally speaking, effects due to Fermi resonance are readily explicable and unlikely to cause serious trouble in structure determinations.

A different reason for the appearance of "additional" fundamentals is the presence in the molecule of one or more atoms having *isotopes*; for if more than one isotope is present in a sample, we shall in fact be dealing with a mixture of different molecular species with different fundamental frequencies. The simplest case is that of a molecule containing only one atom of an element with two isotopes. In the spectra of a sample in which both are present, certain fundamentals will appear "split" into two. With normal instruments, the splitting is unlikely to be resolvable unless the mass ratio of the isotopes is fairly large; and even when this is so, it is unlikely to be observable unless the isotopes are present in comparable abundances.

Striking examples are encountered in the case of boron compounds containing the isotopes ^{10}B and ^{11}B in their natural abundance ratio of about 1:4. We have already mentioned boron trifluoride, for which we gave the observed fundamental frequencies of ^{11}BF$_3$. The infrared and Raman spectra show, also, the corresponding frequencies for ^{10}BF$_3$. The values for the two molecules are compared in Table 5. We see that, unlike the other three, the totally symmetric stretching fundamental is not isotopically split. This fact in itself provides convincing evidence that the structure is planar and not pyramidal, for it proves that the boron nucleus does not take part in the motion of the totally symmetric stretching mode. Were the molecule pyramidal this would not be so. Thus, in this fortunate case the isotopic splitting phenomenon, far from detracting from the usefulness of the spectroscopic method, actually reinforces convincingly the structural conclusion from it.

Table 5. Isotopic Splitting of Fundamentals of BF$_3$

Mode	Activity	Frequency, cm^{-1} ^{11}BF$_3$	^{10}BF$_3$	Isotopic splitting
In-plane deformation	Raman and infrared	480	482	2
Out-of-plane deformation	Infrared only	691	719	28
Totally symmetric stretching	Raman only, polarized	888	888	0
Degenerate stretching	Raman and infrared	1446	1497	51

The situation is less simple when a molecular species contains more than one atom with isotopes in comparable abundances, e.g., the four chlorines in CCl$_4$; for although molecules in which only one isotope is present (e.g., C^{35}Cl$_4$ and C^{37}Cl$_4$) will have the same symmetry, those containing more than one isotope (e.g., C^{35}Cl$_3$^{37}Cl, C^{35}Cl$_2$^{37}Cl$_2$) will have different symmetries and so be subject to different selection rules.[23] In practice, however, the complications will not be serious unless the mass ratio of the isotopes (and consequently the symmetry disturbance) is considerable.

The selection rules apply ideally to free molecules. Wherever practicable, therefore, the spectroscopic evidence should be obtained with gaseous samples at as low a pressure as possible. Often, however, practical considerations make it necessary (especially in the case of Raman spectra) to use liquid samples. The possibility of perturbations due to *molecular interactions* must then be borne in mind. Fortunately, the effects are generally quite small. Indeed, it is remarkable how well, on the whole, the rules of selection and polarization for the free molecule continue to be obeyed by species in pure liquids or in solution. It is true that the values of the fundamental frequencies are somewhat shifted, as compared with the vapor phase; but as far as the selection rules are concerned, the symmetry usually remains effectively the same. Occasionally, however, signs of a failure to conform with the rules of the free molecule may be detected. An example occurs in the Raman spectrum of liquid carbon disulfide, CS_2. The free molecule is symmetrical and linear (like CO_2) and so only the totally symmetric stretching fundamental will be expected to appear. Indeed, it does appear as a very intense polarized line at $\Delta v = 656 \text{ cm}^{-1}$; but the spectrum also contains a relatively feeble line at $\Delta v = 397 \text{ cm}^{-1}$ which, by reference to the infrared, must be identified as the Raman-forbidden bending fundamental. Evidently the selection rules for the free molecule do not hold rigorously for the liquid. That this is due to interactions between CS_2 molecules has been shown by studies of the Raman spectra of solutions of carbon disulfide in an inert solvent. The measured ratio of the intensity of the "forbidden" fundamental to that of the permitted one was found[24] to decrease as the concentration was lowered, i.e., as the average distance between the solute molecules was made greater.

The vibrational spectra will obviously be complicated if molecular interactions occur which are strong enough to produce new chemical species (e.g., by polymerization, complex formation, etc.). Such effects are usually easily recognizable and indeed the spectroscopic method is an excellent one for their detection and characterization.

We may remark in passing that, for the determination of free-molecule symmetry, it is obviously unsatisfactory to rely upon evidence from solid samples, for the solid lattice will not only give rise to new types of vibration but also to new site-symmetry considerations for the component units.

Circumstances Tending to Make the Numbers of Spectral Features Appear Too Small

Here we encounter the most serious practical limitations of the method of determining molecular structure based on application of the vibrational rules of selection and polarization.

For fortuitous mechanical reasons it may be that two permitted fundamentals happen to have very nearly the same frequency. If they are not resolvable, the number of permitted fundamentals will appear to be one too few. As will be seen from later examples, the likelihood of such *fortuitous degeneracy* may be demonstrated by investigation of analogous molecules in which it does not occur; but even so, the circumstances may make it impossible to draw any definite conclusion from an application of selection rules.

More serious, however, is the possible failure to observe permitted features because of inadequacy of available experimental techniques. In this connection we have to remember that the selection rules merely state whether a given fundamental is permitted by symmetry, either in the Raman effect or in infrared absorption, and that they do not tell us anything about how intense a permitted fundamental will in fact be. In certain cases the intensity, though not zero, may be so very low as to be experimentally undetectable. The frequency will then appear to be forbidden. Similarly, the rule of polarization merely states whether a given Raman line will be polarized (i.e., $\rho_n < \frac{6}{7}$), but does not tell us anything about how much less than $\frac{6}{7}$ the ρ_n value of a polarized line will in fact be. In certain cases it may be so near to $\frac{6}{7}$ that the difference may be experimentally undetectable. The line will then appear to be polarized.

Consider the case where the object of an investigation is to decide between two proposed molecular models with different symmetries. As we have seen, the predictions from the rules are more restrictive for the model of higher symmetry, i.e., the numbers of permitted fundamentals and the number of polarized lines will, in general, be smaller than for the model of lower symmetry. It follows that evidence consistent with the model of higher symmetry, though it may be highly suggestive, can never, in principle, be wholly convincing; the possibility must always remain that the additional features permitted for the rival model of lower symmetry may have escaped detection. In other words, the conclusion that the molecule has the higher symmetry must always be based on essentially *negative* observational evidence and consequently however convincing it may seem, it must always be open to some doubt. On the other hand, the conclusion that the molecule has the lower symmetry can be arrived at with a very high degree of confidence, since the evidence in this case will be such as positively to exclude the higher symmetry.

Bearing these all-important considerations in mind, we will now look at some examples of a somewhat less simple kind than those already given.

FURTHER EXAMPLES

"Gallium Dichloride"

The characteristic valency of gallium is three. In the so-called dichloride, of empirical formula $GaCl_2$, it appears to have the valency two. If the molecular formula were $GaCl_2$, the compound would be paramagnetic. In fact it is found to be diamagnetic, and hence it was quite reasonably supposed that the molecule must have the structure $Cl_2Ga-GaCl_2$, involving a bond between the two gallium atoms.

Two models for such a structure deserve consideration. The more likely one would have the pairs of chlorine atoms in the staggered configuration (like the pairs of hydrogens in the allene molecule). The point group for this model is D_{2d}. The alternative would have the pairs of chlorines eclipsed (like the pairs of hydrogens in the ethylene molecule), and the point group would be C_{2v}. For both models the number of distinct fundamentals is found to be nine. As far as the Raman effect is concerned, the predictions from the rules of selection and polarization are as given in Table 6.

In fact, the Raman spectrum of the molten compound was found[25] to consist of only four lines, three of them depolarized and one (the most intense) very highly polarized. This finding obviously does not coincide with the predictions for either of the proposed models (see Table 6). It is conceivable, of course, that the observed spectrum might be seriously incomplete, so that neither model can be definitely excluded; but the degree of incompleteness would be so striking—especially the observation of only one totally symmetric fundamental instead of three—that it becomes at least doubtful whether either of the proposed structures can be the correct one.

This looks like a very unfortunate case; but further considerations show that in fact it is a singularly fortunate one. The observed frequen-

Table 6. Predictions for Fundamentals of Ga_2Cl_4

Model	Number permitted in Raman effect	Number of polarized Raman lines
Staggered (D_{2d})	9	3
Eclipsed (planar—C_{2v})	6	3

cies and states of polarization of the lines are as follows:

> $115\ cm^{-1}$, depolarized,
> $153\ cm^{-1}$, depolarized,
> $346\ cm^{-1}$, very strong and highly polarized,
> $380\ cm^{-1}$, depolarized.

For the regular tetrahedral $GaCl_4^-$ ion in aqueous solution the corresponding findings[17] were previously known to be the following:

> $114\ cm^{-1}$, depolarized
> $149\ cm^{-1}$, depolarized
> $346\ cm^{-1}$, very strong and highly polarized
> $386\ cm^{-1}$, depolarized.

Making allowances for minor frequency changes due to change of environment (molten compound as compared with aqueous solution), the close similarity between the two spectra is very striking. Indeed, they are practically identical. The evidence is thus very convincing that "gallium dichloride" must have the ionic structure $[Ga^+][GaCl_4^-]$, and that the apparent valency of two is due to the presence of one Ga atom of valency one and a second of valency three. This structure, first deduced from the Raman spectrum, was later fully confirmed by other methods.

As already pointed out, this case is a singularly lucky one, for it must indeed be rare that, after investigating the question of the structure of a new compound, it should be discovered that the key to the correct answer was known before starting.

Oxygen Disilyl (Disiloxane)

The example of Ga_2Cl_4 given above represents a success for our method, albeit a success made possible by a special circumstance. We come now to a case of failure.

The silyl group, $-SiH_3$, is analogous to the methyl group, $-CH_3$, but may (as we shall see) give rise to interesting stereochemical differences. It is known from electron-diffraction evidence that in oxygen dimethyl (dimethyl ether) the two O–C bonds make an angle of about 111° with one another. The COC skeleton thus belongs to the point group C_{2v} (see Table 1). This is in accordance with expectation for sp^3 type hybridization and a valency shell for the oxygen atom containing two lone pairs of electrons. In oxygen disilyl the skeleton might be of the same shape; but because the silicon atoms (unlike the carbon atoms in oxygen dimethyl) possess vacant d-orbitals, the possibility must be envisaged that these might accept the lone pairs from the oxygen forming $p \rightarrow d$ π-bonds. In consequence, the hybridization of the central atom might well be changed to the linear sp type. The skeleton (formally written as $Si \leftharpoonup O \rightleftharpoons Si$) would then become linear and belong to the point group $D_{\infty h}$ (see Table 1).

A decision between the two models can, in principle, be made by the vibrational spectroscopic method, in the application of which we may reasonably expect that it will be sufficient to neglect the light hydrogen atoms and to consider only the heavier skeleton as if it were a triatomic species.

Before proceeding to the disilyl, we may confirm the validity of this expectation by considering the skeletal spectra of typical *dimethyls* whose skeletal shapes are known. Examples with linear skeletons are rather few: probably the most completely investigated is mercury dimethyl, $Hg(CH_3)_2$. In accordance with the predictions for symmetry $D_{\infty h}$ (see Table 1), only two of the three skeletal fundamentals are found[26,27] in the infrared (550 and 153 cm^{-1}), and the remaining one (symmetric stretching, $\Delta v = 575$ cm^{-1}) appears only in the Raman spectrum[26] where its line is intense and polarized. There is a minor complication in that the Raman spectrum of the liquid also shows[27] the lower of the two infrared-permitted fundamentals (skeletal bendings), which is Raman-forbidden according to the free-molecule selection rules. The line is very feeble, and the case seems to be analogous to that of liquid carbon disulfide, the feeble appearance of the "forbidden" deformation being due to molecular interactions in the liquid.

An example of a dimethyl with a bent skeleton is oxygen dimethyl (dimethyl ether). In accordance with the predictions for the point group C_{2v} (see Table 1), all three skeletal fundamentals appear[28] in both the Raman spectrum [$\Delta v = 1104$, 922 (polarized), and 428 (polarized) cm^{-1}] and the infrared. Thus altogether the findings for dimethyls give confidence that the shape of the skeleton can be diagnosed from a consideration of the rules of selection and polarization for the skeletal frequencies alone. Where it is not at once obvious from their frequency ranges which are the skeletal fundamentals and which are those principally involving motions of the hydrogen atoms, the latter can be readily identified by investigating the corresponding completely deuterated compound. Those frequencies involving mainly motions of the hydrogen atoms will be shifted downward very considerably on deuteration, whereas the skeletal frequencies will be relatively very little affected.

In the encouraging light of the findings for dimethyls we may now go on to consider the very interesting case of *oxygen disilyl*. The skeletal Raman spectrum was found[29] to consist of just one intense and polarized line ($\Delta v = 606\ cm^{-1}$), obviously the symmetric stretching fundamental v_1. This frequency was absent from the infrared absorption spectrum, which showed, instead, a strong skeletal fundamental at $1107\ cm^{-1}$, obviously due to asymmetric stretching, v_2. These observations are consistent with the linear model. The third skeletal frequency (bending, v_3) is also permitted in the infrared, and the fact that it was not observed in the quoted work was accounted for on the reasonable grounds that it probably lies below the lower limit ($280\ cm^{-1}$) of the range accessible to the spectrometer used. It was later located[30] at approximately $68\ cm^{-1}$.

Bearing in mind the precedents from dimethyl compounds, it was natural to conclude that the Si–O–Si skeleton of oxygen disilyl is linear. However, despite the apparently convincing spectroscopic evidence, this conclusion turned out to be wrong. Subsequent careful electron-diffraction investigation[31] showed, in fact, that the angle between the two O–Si bonds, far from being 180°, is 144°. This is indeed a large bond angle as compared with 111° in oxygen dimethyl, but it is remarkable that the vibrational spectra should speak so definitely (and so deceptively) in favor of the linear model. It is all the more remarkable in view of the essential physical difference between the mechanisms of Raman scattering and infrared absorption; for we should hardly have anticipated that two such different processes would, so to speak, have conspired together to support one another in deceiving the observer.

Note that this is a case where the spectroscopic evidence is essentially negative in character: support for the conclusion of linearity is the *absence* of v_2 and v_3 from the observed Raman spectrum. This raises the question of the observable effects upon the spectra if, starting with a strictly linear model, we imagine the bond angle to be gradually decreased from 180°. In principle, of course, even the smallest departure from the linearity will cause v_2 and v_3 to be permitted in the Raman effect and v_1 to be permitted in the infrared, though we shall expect that, in practice, their intensities will gradually increase from zero as the bond angle is progressively made smaller. In the absence of a satisfactory theory of intensities for either type of spectrum it is impossible to predict what degree of bending will be required in order to bring them into the range of experimental detectability. At lesser degrees of bending it will appear to the observer that the selection rules for the linear model are still being obeyed.

The empirical findings are represented diagrammatically in Fig. 1, where the convenient term "pseudolinear" is used to describe a model which is in fact bent but which gives spectra that simulate obedience to the linear selection rules.

As we have seen, the observation of only one Raman line, v_1, contributed to the original erroneous conclusion that the skeleton of oxygen disilyl is linear. It is, therefore, especially interesting to note that Raman evidence on this line was later adduced to prove that the skeleton must, in fact, be nonlinear. This was done by studying the effect of replacing the central oxygen atom (^{16}O) by its isotope ^{18}O. If the skeleton were linear, the central atom would not partake in the motion of the symmetric stretching mode and so v_1 would remain unshifted; whereas with a nonlinear skeleton this atom must necessarily move, and so v_1 must suffer an isotopic shift. (Compare the case of $^{10}BF_3$ and $^{11}BF_3$.) In fact, a small but definite shift was observed.[32]

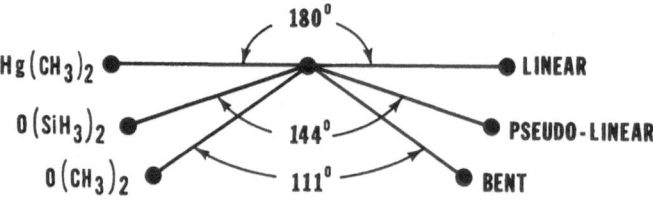

Fig. 1. Skeletal selection rules for molecules of type XY_2.

Sulfur and Selenium Disilyls

The difficulty of pseudolinearity, encountered with oxygen disilyl, does not occur with the sulfur and selenium analogues. For $S(SiH_3)_2$ all three skeletal fundamentals are observed[33] in the Raman spectrum (Δv = 159, 480, and 508 cm^{-1}), from which it is certain that the skeleton must be bent. We note that, in consequence of the increased mass of the central atom, the two skeletal stretching frequencies are very much closer together (difference of only 28 cm^{-1}) as compared with those of the oxygen analogue (difference of approximately 500 cm^{-1}, see above). It is not surprising, therefore, that for $Se(SiH_3)_2$ the two should have become so nearly equal as to be unresolvable in practice. In fact, the Raman spectrum of $Se(SiH_3)_2$ shows,[33] in addition to the bending fundamental (Δv = 130 cm^{-1}), just one line (Δv = 388 cm^{-1}) which doubtless consists of both stretching frequencies in superposition. There can be no doubt from this (positive) evidence that the disilyls of both sulfur and selenium have nonlinear skeletons. This conclusion is borne out by the infrared absorption spectra.[33] A subsequent electron-diffraction study[34] of the sulfur compound showed that the Si–S–Si angle is 97°.

Nitrogen Trisilyl (Trisilylamine)

The idea that the skeleton of oxygen disilyl might be linear arose from consideration of the capacity of the Si atom to form $p \rightarrow d \pi$-bonds. Although linearity is not in fact attained, the abnormally large bond angle in oxygen disilyl indicates that π-bond formation is indeed producing an appreciable stereochemical effect in the expected direction. Now in nitrogen trisilyl the number of Si atoms is greater (three instead of two), and the number of lone pairs of electrons on the central atom is smaller (one instead of two). It is, therefore, likely that the effects of π-bonding will be more marked, and with bond resonance the possibility of a planar skeleton (sp^2 hybridization on the N atom), as contrasted with the pyramidal skeleton of $N(CH_3)_3$ (sp^3 hybridization), cannot lightly be dismissed.

Before discussing the vibrational spectroscopic evidence for nitrogen trisilyl, let us first look at typical examples of *trimethyls* in order to see whether they give support for the method of diagnosing skeletal structure from the selection rules for skeletal fundamentals alone. There is no doubt that boron trimethyl has a planar skeleton (point group D_{3h}). The observed Raman and infrared spectra[35] are in full agreement with the predictions (see Table 3) from the D_{3h} selection rules. In particular, the totally symmetric skeletal stretching frequency appears as a strong polarized Raman line ($\Delta v = 680\,cm^{-1}$), but is absent from the infrared; the out-of-plane skeletal deformation appears in the infrared (336, 345 cm^{-1}, isotopic splitting), but is absent from the Raman spectrum. As an example of a trimethyl which certainly has a pyramidal skeleton (point group C_{3v}) we may take nitrogen trimethyl (trimethylamine). Again in accordance with prediction (see Table 3), all four skeletal fundamentals are found, both in the Raman[36] and the infrared[37] spectrum.

Encouraged by the above findings, we may now proceed to the case of *nitrogen trisilyl* with a considerable measure of confidence that we shall be able to diagnose its skeletal shape from a consideration of its skeletal fundamentals alone. The experimental evidence for this molecule, and its completely deuterated analogue, shows[38] that it conforms with the predictions for a planar skeleton. In particular, the totally symmetric skeletal stretching frequency is observed in the Raman spectrum as an intense, polarized line ($\Delta v = 496\,cm^{-1}$), but seems to be absent from the infrared. Also, there is no evidence of the appearance of the out-of-plane skeletal deformation frequency in the Raman spectrum. (Permitted in the infrared, it presumably lies below the range of the spectrometer used.) On this evidence it was concluded that the skeleton is planar, i.e., that the π-bonding capacity of the Si atoms strikingly changes the geometry as compared with the trimethyl. However, it must be admitted that here, as always where the more highly symmetrical of two models is deemed to be the correct one, the evidence is essentially negative in kind. In this particular case, however, the conclusion from vibrational spectroscopy is supported by electron diffraction evidence. Nevertheless, it would be prudent to claim only that the NSi_3 skeleton is pseudoplanar; though in view of the electron-diffraction support, the emphasis in this particular case should doubtless be on *planar* rather than on pseudo.

Trisilyls of Other Group V Elements

Because of the planarity (or at least near planarity) of $N(SiH_3)_3$, it was interesting to inquire whether the phosphorus, arsenic, and antimony analogues also show this remarkable stereochemical phenomenon. The vibrational spectroscopic method is clearly a relevant one.

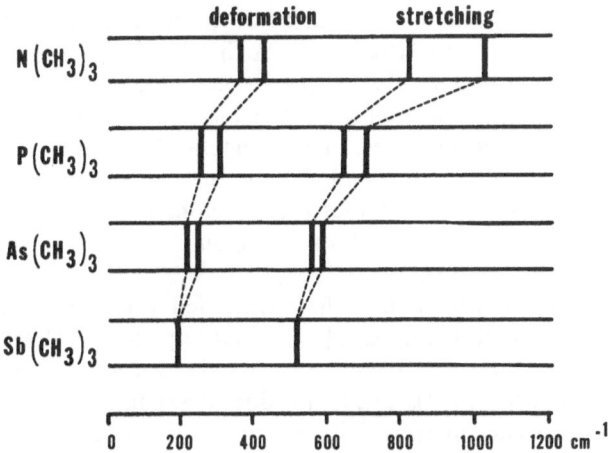

Fig. 2. Skeletal fundamentals of Group V trimethyls.

We may first note that, in view of the effects of increasing the mass of the central atom in disilyls (see above), certain limitations are to be anticipated for the trisilyls on account of near degeneracies of fundamental skeletal frequencies. Indeed, limitations of this kind impose themselves[39] in the case of the *trimethyls* of Group V elements (all pyramidal). This is shown diagrammatically in Fig. 2. As the mass of the central atom increases, we see that the two skeletal stretching frequencies draw nearer together, until in $Sb(CH_3)_3$ they are practically unresolvable. The same is true also of the two skeletal deformation frequencies. Thus, for $Sb(CH_3)_3$ [and the same is true for $Bi(CH_3)_3$], although all four fundamentals are permitted in both the Raman and infrared spectra, it is clear why only two frequencies are in fact observed—each doubtless consisting of two fundamentals in superposition.

A similar state of affairs may be anticipated for the trisilyls; and indeed it is found for $Sb(SiH_3)_3$. Figure 3 shows the progressive changes of the two observed skeletal stretching frequencies as the mass of the central Group V atom increases. (Compare Fig. 2 for the trimethyls.) In $Sb(SiH_3)_3$ the two have become practically coincident. A similar behavior is to be expected for the two skeletal deformations (compare Fig. 2).

The observed spectra of the trisilyls of phosphorus[40] and arsenic[41] are both of the kind already described for $N(SiH_3)_3$. Specially notable is that in both cases the totally symmetric skeletal stretching frequency appears in the Raman spectrum as a strong, polarized line ($\Delta v =$ 414 cm^{-1} for PSi$_3$ and 346 cm^{-1} for AsSi$_3$), but is not observed in the infrared. According to the spectroscopic evidence, both molecules appear to obey the selection rules for a planar skeleton; the observers wisely pointed out, however, that the possibility remained open that the shape might be merely pseudoplanar. In fact, a subsequent electron-diffraction study[42] showed that the skeletons of both molecules are pyramidal. These cases are thus analogous to that of oxygen disilyl (but see further below).

As to antimony trisilyl, the anticipated presence of fortuitous degeneracies made it impossible to draw any structural conclusions at all. The Raman spectrum[41] shows only two skeletal frequencies ($\Delta v = 309$ and 99 cm^{-1}). The higher doubtless consists of both stretching frequencies in superposition (see Fig. 3), but it is not possible to say whether or not the lower contains both deformation frequencies. Similarly, in the infrared it is not possible to say whether or not the single higher-frequency feature contains the totally symmetric stretching fundamental as well as the degenerate stretching one.

As with the dimethyl and disilyl compounds of Group VI elements (see Fig. 1), the question arises here as to how great the departure must be from a model of higher symmetry before it becomes experimentally possible to detect the departure from obedience to the selection rules for that model. The factors determining intensities are not well enough understood to allow of any theoretical answer. Indeed the

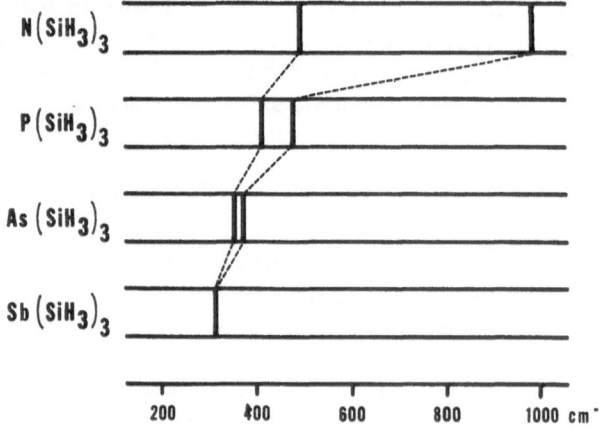

Fig. 3. Skeletal stretching fundamentals of Group V trisilyls.

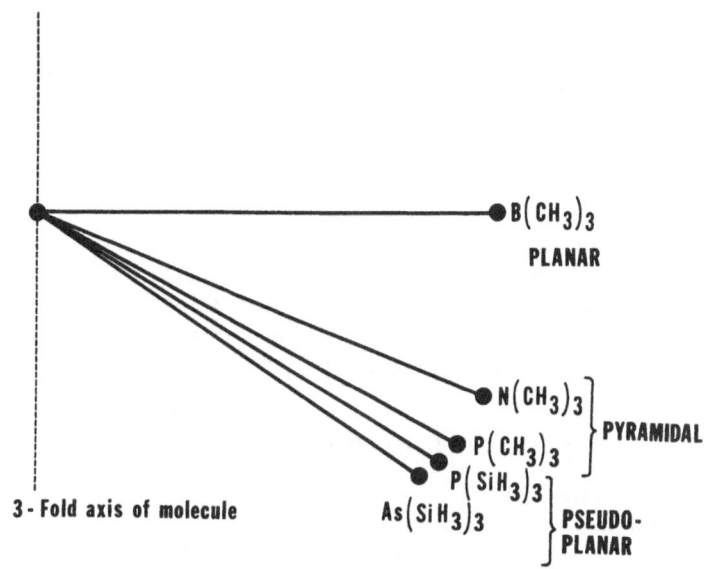

Fig. 4. Skeletal selection rules for molecules of type XY_3.

empirical findings for XY_3-type skeletons show that the question is less straightforward than might be supposed. The planar selection rules are obeyed by $B(CH_3)_3$ which certainly has a planar skeleton (bond angles, 120°), and also by $N(SiH_3)_3$ where the skeleton is doubtless planar or very nearly so. The pyramidal rules are obeyed by $N(CH_3)_3$ and $P(CH_3)_3$ whose bond angles (as given by electron diffraction) are, respectively, 108° and 110°. On the other hand, as we have seen $P(SiH_3)_3$ and $As(SiH_3)_3$ are both pyramidal, yet simulate obedience to the planar rules, i.e., must be described as pseudoplanar. By analogy with the pseudolinear oxygen disilyl (see Fig. 1) it would be natural to expect that the bond angles of $P(SiH_3)_3$ and $As(SiH_3)_3$ would be intermediate between those of the planar $B(CH_3)_3$ and the pyramidal $N(CH_3)_3$, i.e., between 120° and 108°. Surprisingly, however, the electron-diffraction values[42] are approximately $96\frac{1}{2}°$ for $P(SiH_3)_3$ and 94° for $As(SiH_3)_3$. Both are *smaller* than for $N(CH_3)_3$, i.e., represent a *greater* departure from planarity.

The situation is represented diagrammatically in Fig. 4, in which, for clarity of presentation, the departures from planarity are shown not by the bond angles but by the angles between each bond and the threefold molecular axis. It is clear that the onset of experimentally observable obedience to the pyramidal selection rules is not determined solely by the geometrical extent of departure from the planar configuration, but must also depend upon the specific nature of the atoms involved. Thus, the simpler impression given by Fig. 1 is illusory.

Phosphorus Trigermyl

The doubts and difficulties encountered with the trisilyls do not occur for phosphorus trigermyl, $P(GeH_3)_3$. Four distinct skeletal fundamentals ($\Delta v = 88$, 112, 322, and 366 cm^{-1}) are observed[43] in the Raman spectrum, the second and third being polarized. This positively proves that the skeleton is pyramidal. Supporting evidence is supplied by the infrared spectrum.[43]

CONCLUDING COMMENTS

Many further examples could be adduced, but it is hoped that the foregoing selection will have given a general idea of the power and limitations of the method. In some cases, observation of one kind of vibrational spectrum only (either Raman or infrared) may seem to give sufficient evidence for a structural determination; but in view of the essentially different and complementary natures of the two, it is very advisable (wherever practicable) to investigate the other kind of spectrum also. Even when this is done, the conclusion as to point group may still be open to doubt, especially when the spectroscopic evidence appears to favor the higher of two proposed symmetries. Independent evidence should always be sought by other methods, such as electron diffraction, microwave spectroscopy, nuclear magnetic resonance, etc. As we have seen, vibrational spectroscopy is by no means all-powerful, and should therefore be regarded as just one contributory method among others—often a very useful one.

If a good deal of this chapter reads like a cautionary tale, that is because from extended experience in this field the writer has learned the necessity for caution.

REFERENCES

1. L. A. Woodward, *Phil. Mag.* **18**: 823 (1934).
2. H. Braune and G. Engelbrecht, *Z. Physik. Chem.* **B19**: 303 (1932).
3. W. Klemperer and L. Lindemann, *J. Chem. Phys.* **25**: 397 (1956).
4. S. Brahms and J.-P. Mathieu, *Compt. Rend.* **251**: 938 (1960).
5. R. D. Shelton, A. H. Nielsen, and W. H. Fletcher, *J. Chem. Phys.* **21**: 2178 (1953).

6. E. F. Barker, *Rev. Mod. Phys.* **14**: 198 (1942).
7. H. H. Hyman, *Science* **145**: 773 (1964).
8. E. K. Plyler and E. F. Bender, *Phys. Rev.* **38**: 1827 (1931); *ibid.* **41**: 369 (1932).
9. J. Cabannes and A. Rousset, *J. Phys. Radium* (8) **1**: 210 (1940).
10. D. M. Gage and E. F. Barker, *J. Phys. Chem.* **7**: 455 (1939).
11. T. F. Anderson, E. N. Lassetre, and D. M. Yost, *J. Chem. Phys.* **4**: 703 (1936).
12. J. R. Nielsen and N. E. Ward, *J. Chem. Phys.* **10**: 81 (1942).
13. P. W. Davis and R. A. Oetjen, *J. Mol. Spectry.* **2**: 253 (1958).
14. H. H. Claassen, B. Weinstock, and J. G. Malm, *J. Chem. Phys.* **28**: 285 (1958).
15. D. F. Smith, *J. Chem. Phys.* **21**: 609 (1953).
16. M.-L. Delwaulle, M. F. Francois, and M. Delhaye-Buisset, *J. Phys. Radium* **15**: 206 (1954).
17. L. A. Woodward and A. A. Nord, *J. Chem. Soc.* 3721 (1956).
18. D. M. Adams and H. A. Gebbie, *Spectrochim. Acta* **19**: 925 (1963).
19. H. Stammreich and R. Forneris, *Spectrochim. Acta* **16**: 363 (1960).
20. H. H. Claassen, C. L. Chernick, and J. G. Malm, *J. Am. Chem. Soc.* **85**: 1927 (1963).
21. R. E. Dodd, L. A. Woodward, and H. L. Roberts, *Trans. Faraday Soc.* **52**: 1052 (1956).
22. A. Langseth and J. R. Nielsen, *Phys. Rev.* **46**: 1057 (1934).
23. C. K. Wu and G. B. B. M. Sutherland, *J. Chem. Phys.* **6**: 114 (1938).
24. J. C. Evans and H. J. Bernstein, *Can. J. Chem.* **34**: 1127 (1956).
25. L. A. Woodward, G. Garton, and H. L. Roberts, *J. Chem. Soc.* 3723 (1956).
26. H. G. Gutowsky, *J. Chem. Phys.* **17**: 128 (1949).
27. L. A. Woodward, *Spectrochim. Acta* **19**: 1963 (1963).
28. Y. Kanazawa and K. Nukada, *Bull. Chem. Soc. Japan* **35**: 612 (1962).
29. R. C. Lord, D. W. Robinson, and W. C. Schumb, *J. Am. Chem. Soc.* **78**: 1327 (1956).
30. D. W. Robinson, W. J. Lafferty, J. R. Aronson, J. R. Durig, and R. C. Lord, *J. Chem. Phys.* **35**: 2245 (1961).
31. A. Almenningen, O. Bastiansen, V. Ewing, K. Hedberg, and M. Traettberg, *Acta Chem. Scand.* **17**: 2455 (1963).
32. D. C. McKean, R. Taylor, and L. A. Woodward, *Proc. Chem. Soc.* 321 (1959).
33. E. A. V. Ebsworth, R. Taylor, and L. A. Woodward, *Trans. Faraday Soc.* **55**: 211 (1959).
34. A. Almenningen, K. Hedberg, and R. Seip, *Acta Chem. Scand.* **17**: 2264 (1963).
35. L. A. Woodward, J. R. Hall, R. N. Nixon, and N. Sheppard, *Spectrochim. Acta* **15**: 249 (1959).
36. R. Ananthrakrishnan, *Proc. Indian Acad. Sci.* **A4**: 204 (1936).
37. J. R. Barcelo and J. Bellanato, *Spectrochim Acta* **8**: 27 (1956).
38. E. A. V. Ebsworth, J. R. Hall, M. J. Mackillop, D. C. McKean, N. Sheppard, and L. A. Woodward, *Spectrochim. Acta* **13**: 202 (1958).
39. E. J. Rosenbaum, D. J. Ruben, and C. R. Sandberg, *J. Chem. Phys.* **8**: 366 (1940).
40. G. Davidson, E. A. V. Ebsworth, G. M. Sheldrick, and L. A. Woodward, *Spectrochim. Acta* **22**: 67 (1966).
41. G. Davidson, L. A. Woodward, E. A. V. Ebsworth, and G. M. Sheldrick, *Spectrochim. Acta* **23A**: 2069 (1967).
42. B. Beagley, A. G. Robinette, and G. M. Sheldrick, *Chem. Commun.*: 601 (1967); *J. Chem. Soc.*: A, 3002 and 3006 (1968).
43. S. Cradock, G. Davidson, E. A. V. Ebsworth, and L. A. Woodward, *Chem. Commun.*: 515 (1965).

Chapter 2

Developments in the Theories of Vibrational Raman Intensities*

J. Tang† and A. C. Albrecht

Department of Chemistry
Cornell University
Ithaca, New York

INTRODUCTION

The theory of light scattering from an individual particle involving the "summation-over-states" problem of second-order perturbation theory, was offered first in 1925 by Kramers and Heisenberg[1] by analogy to the classical theory of dispersion. The same expression was verified later by Dirac[2] based on the radiation-field theory. And from this expression Van Vleck[3] derived the basic selection rules for the vibrational Raman effects in molecules. For the nonresonance case, special simplifications are possible. When the frequency of the incident light is far from the resonance region, and the initial and final states of the molecule are both in the ground electronic state, it was shown by Placzek[4] that the intensity of molecular Raman scattering arises from the dependence of the ground-state polarizability on nuclear vibrations. This so-called polarizability theory of Placzek has led to the bond polarizability theory commonly used by chemists to analyze vibrational Raman spectra in terms of ground-state properties of the molecules. (The bond polarizability theory is reviewed in the first volume of this book.) The object of the present chapter is to review the recent developments of Raman intensity theories, beginning, more or less, at the stage subsequent to the material treated in the first volume and continuing to the most recent work which has appeared in the literature (1960–1968).

There are two main approaches in the recent studies of the theories of vibrational Raman intensities. One approach is to study the

*This work has been assisted by a grant from the National Science Foundation.
†Present address: Department of Chemistry, University of California at Irvine.

Kramers–Heisenberg–Dirac dispersion expression as a problem of vibronic spectroscopy. In this approach each term in the dispersion expression represents a contribution from the transition between the initial (or final) state of the molecule to an excited virtual vibronic state. Thus, knowledge of the properties of the excited vibronic states is closely tied to the vibrational Raman intensities exhibited by the ground electronic state. And, thus, one can use Raman spectroscopy as a tool for studying the properties of the excited electronic states, as a complementary method of vibronic spectroscopy. This was pointed out some time ago in a study by Albrecht,[5] where an explicit relationship was found between the Raman intensities of the ground-state vibrations and the vibronic intensities of allowed transitions of a molecule. At the same time, Krushinskii and Shorygin[6–8] modified the semiclassical theory of Shorygin[9,10] for totally symmetric modes (see Chapter 6 of the first volume of this book). The so-called "quantum model" thus obtained bears a certain relationship to Albrecht's[5] expression. Savin[11–13] and Verlan[14,15] subsequently developed a more general expression which, in essence, contains the substance of both approaches. All these treatments will be discussed in this chapter after introducing a small modification to the familiar dispersion expression for atoms (Kramers–Heisenberg–Dirac) so that it may apply to molecules having a permanent dipole moment.

The second major approach to Raman scattering theory is to study the dispersion expression as a problem of the nuclear coordinate dependence of the ground-state polarizability. In this approach, the sum over excited states in the dispersion expression is closed formally. Now the explicit dependence of the excited molecular states disappears in the new expression, and only ground electronic state wave functions are seen. Since the exact closure of the sum cannot be evaluated in any ordinary case, various approximations are employed. One approximation generally used appears in the treatment of ground-state polarizabilities using variation methods. The nuclear coordinate dependence of a simple approximate expression for polarizability then leads to a calculation of Raman intensities. An especially simple expression for the intensity of a given totally symmetric mode was derived by Long and Plane,[16] and a similar one by Lippincott and Nagarajan,[17] using a delta-function-potential model. A different treatment of the closure of the sum by Tang and Albrecht[18] results in another expression for the Raman intensities, which can presumably be applied to all normal modes. These "ground-state" expressions are presented and discussed in the second part of this chapter.

VIBRONIC EXPANSION APPROACH

Basic Dispersion Equation for Molecules

The total intensity of a Raman line after averaging over all orientations of the particle, is given by [see equation (17) in Chapter 6 of Volume 1]

$$I_{m,n} = \frac{2^3\pi}{3^2c^4} I_0\omega^4 \sum_{\rho\sigma} |(\alpha_{\rho\sigma})_{m,n}|^2 \tag{1}$$

where I_0 is the intensity of the incident light, ω is the frequency of the scattered light ($\omega = 2\pi\nu$), $\alpha_{\rho\sigma}$ is the ρ,σth component of the polarizability tensor, m is the initial state, and n is the final state.

Now, $(\alpha_{\rho\sigma})_{m,n}$ coming from the induced transition moment $P_{m,n} = E \cdot (\alpha)_{m,n}$, which is derived by second-order perturbation theory, is

$$(\alpha_{\rho\sigma})_{m,n} = \frac{(e/\mu_e)^2}{\omega\omega_0} \sum_e \left[\frac{\langle m|P_\sigma|e\rangle\langle e|P_\rho|n\rangle}{E_e - E_m - E_0} + \frac{\langle m|P_\rho|e\rangle\langle e|P_\sigma|n\rangle}{E_e - E_n + E_0} \right] \tag{2}$$

where E_0 is equal to $\hbar\omega_0$ and is the energy of the incident light which is in the visible or ultraviolet region, e/μ_e is the ratio of charge to mass of the electron, P_σ is equal to $\sum_k (p_k)_\sigma$ and is the scalar amplitude of the σth component of the many-electron momentum operator, and $|e\rangle$ is an intermediate state. Notice that the operators act on electrons only, because nuclear motion does not respond, significantly, to light in the visible region and is, thus, not included in the perturbation (however, see Appendix A). The momentum operator is related to the dipole-moment operator in a simple way. The familiar dipole version of equation (2) is as follows:

$$(\alpha_{\rho\sigma})_{m,n} = \sum_e{}' \left[\frac{\langle m|R_\sigma|e\rangle\langle e|R_\rho|n\rangle}{E_e - E_m - E_0} + \frac{\langle m|R_\rho|e\rangle\langle e|R_\sigma|n\rangle}{E_e - E_n + E_0} \right] \tag{3}$$

[equation (8) in Chapter 6 of Volume 1], where R_σ and R_ρ are the electronic dipole-moment operators, e.g., $R_\sigma = - \sum_k e(r_k)_\sigma$, with $(r_k)_\sigma$ being the σth component of the position vector of the kth electron. This expression was derived by Dirac for atoms.[2] In applying it to molecules, however, one must be more careful and more explicit about the meaning of the expression. It is shown in Appendix A of this chapter that excluding the initial and the final states in the summation [whence the prime notation in equation (3)], becomes crucial whenever a molecule possesses a dipole moment. Furthermore, one can also exclude all the vibrational substates of the ground electronic level whenever it is possible to represent such a state by a product wave function, e.g.,

$$\Psi_m = \theta_g(r, Q)\phi_i(Q), \qquad (\text{or } |m\rangle = |g\rangle|i\rangle)$$

One can alternatively write

$$(\alpha_{\rho\sigma})_{m,n} = \sum_e \left[\frac{\langle m|M_\sigma|e\rangle \langle e|M_\rho|n\rangle}{E_e - E_m - E_0} + \frac{\langle m|M_\rho|e\rangle \langle e|M_\sigma|n\rangle}{E_e - E_n + E_0} \right] \qquad (4)$$

where the summation is over *all* vibronic states, and the operators M_σ and M_ρ are defined by $M_\sigma = R_\sigma - (g|R_\sigma|g)$. Equation (4) becomes exactly the same expression as that derived by Dirac, and Kramers and Heisenberg, when the integral over electronic coordinates, $(g|R_\sigma|g)$, vanishes as in case of atoms and also for molecules with no dipole moment.

Adiabatic Approximation

To treat Raman intensity as a problem of vibronic interactions, we first assume the adiabatic approximation for all wave functions involved, and then expand the electronic parts around the ground-state equilibrium configurations of the nuclei by a Herzberg–Teller[19] series to the first order.

Let us choose the following notation for the adiabatic approximation:

$$|m\rangle = |g\rangle\|i\rangle \quad \text{or} \quad \Psi_m(r, Q) = \theta_g(r, Q)\phi_i^g(Q)$$

$$|n\rangle = |g\rangle\|j\rangle \quad \text{or} \quad \Psi_n(r, Q) = \theta_g(r, Q)\phi_j^g(Q) \qquad (5a)$$

$$|e\rangle = |e\rangle\|v\rangle \quad \text{or} \quad \Psi_e(r, Q) = \theta_e(r, Q)\phi_v^g(Q)$$

Thus, equation (3) appears as

$$(\alpha_{\rho\sigma})_{gi,gj} = \sum_{e,v}{}' \left[\frac{\langle i\|(g|R_\sigma|e)\|v\rangle \langle v\|(e|R_\rho|g)\|j\rangle}{E_{ev} - E_{gi} - E_0} \right.$$
$$\left. + \frac{\langle i\|(g|R_\rho|e)\|v\rangle \langle v\|(e|R_\sigma|g)\|j\rangle}{E_{ev} - E_{gj} + E_0} \right] \qquad (5b)$$

We choose to use equation (3) as the starting point instead of equation (2) or equation (4). All the approximations applied to equation

(3) can be applied to equation (2) and equation (4) in an analogous way. Equation (4) is more convenient in the second section of this chapter when closure of the summation over excited states is considered. In this section, it is more convenient to use equation (3) because most of the literature in the vibronic expansion approach does so. Exceptions are in the papers by Kondilenko et al.[20] who chose to use equation (2), having the momentum operators instead of dipole-moment operators.

Herzberg–Teller Expansion. The Herzberg–Teller expansion[19] is the formal expansion of an electronic wave function (under the adiabatic approximation) in a Taylor's series of displacement of the nuclear coordinates from some equilibrium position in the ground state. The coefficients of the linear terms are identified with the vibronic coupling operator $h_a^0 = (\partial \mathcal{H}/\partial Q_a)^0$ (for normal mode a) through the perturbation method. Thus,

$$|e) = |e^0) + \sum_a \sum_{s \neq e}{}' \frac{(h_a)_{es}^0 \cdot \Delta Q_a}{E_e^0 - E_s^0} |s^0) \tag{6}$$

and $|g)$ becomes

$$|g) = |g^0) + \sum_a \sum_{t \neq g} \frac{(h_a)_{gt}^0 \cdot \Delta Q_a}{E_g^0 - E_t^0} |t^0) \tag{7}$$

Upon applying the Herzberg–Teller expansion to the electronic wave functions in equation (5), we obtain

$$(\alpha_{\rho\sigma})_{gi,gj} = A + B + C \tag{8}$$

$$A = \sum_{e \neq g}{}' \sum_v \left[\frac{(g^0|R_\sigma|e^0)(e^0|R_\rho|g^0)}{E_{ev} - E_{gi} - E_0} + \frac{(g^0|R_\rho|e^0)(e^0|R_\sigma|g^0)}{E_{ev} - E_{gj} + E_0} \right] \langle i\|v\rangle \langle v\| j\rangle$$

$$B = \sum_{e \neq g}{}' \sum_v \sum_{s \neq e}{}' \sum_a$$

$$\times \left[\frac{(g^0|R_\sigma|e^0)(e^0|h_a|s^0)(s^0|R_\rho|g^0)}{E_{ev} - E_{gi} - E_0} + \frac{(g^0|R_\rho|e^0)(e^0|h_a|s^0)(s^0|R_\sigma|g^0)}{E_{ev} - E_{gj} + E_0} \right]$$

$$\times \langle i\|v\rangle \langle v\|Q_a\| j\rangle / (E_e^0 - E_s^0)$$

$$+ \left[\frac{(g^0|R_\sigma|s^0)(s^0|h_a|e^0)(e^0|R_\rho|g^0)}{E_{ev} - E_{gi} - E_0} + \frac{(g^0|R_\rho|s^0)(s^0|h_a|e^0)(e^0|R_\sigma|g^0)}{E_{ev} - E_{gj} + E_0} \right]$$

$$\times \langle i\|Q_a\|v\rangle \langle v\|j\rangle / (E_e^0 - E_s^0)$$

$$C = \sum_{e \neq g}' \sum_{t \neq g}' \sum_v \sum_a$$

$$\times \left[\left\{ \frac{(g^0|h_a|t^0)(t^0|R_\sigma|e^0)(e^0|R_\rho|g^0)}{E_{ev} - E_{gi} - E_0} + \frac{(g^0|h_a|t^0)(t^0|R_\rho|e^0)(e^0|R_\sigma|g^0)}{E_{ev} - E_{gj} + E_0} \right\} \right.$$

$$\times \frac{\langle i\|v\rangle \langle v\|Q_a\|j\rangle}{E_g^0 - E_t^0}$$

$$+ \left\{ \frac{(g^0|R_\sigma|e^0)(e^0|R_\rho|t^0)(t^0|h_a|g^0)}{E_{ev} - E_{gi} - E_0} + \frac{(g^0|R_\rho|e^0)(e^0|R_\sigma|t^0)(t^0|h_a|g^0)}{E_{ev} - E_{gj} + E_0} \right\}$$

$$\left. \times \frac{\langle i\|Q_a\|v\rangle \langle v\|j\rangle}{E_g^0 - E_t^0} \right]$$

Having presented this general vibronic representation of the Raman intensity theory, it is of interest to discuss next the various investigations within this context.

An Early Approximate Form of the Vibronic Theory. In an early examination[5] of the vibronic theory for the nonresonance Raman case two approximations are involved. The first treats the energy denominator as constant over v for energy e so that the sum over v is a simple matrix multiplication expansion for every e. After doing the sum over v, one discovers that term A contributes only to Rayleigh scattering but not to Raman. The second approximation assumes $|g) = |g^0)$ in equation (7). This is the usual assumption in vibronic transitions because the denominators $(E_g^0 - E_t^0)$ in the expansion of the ground state are usually much larger in magnitude than $(E_e^0 - E_s^0)$ in the expansion of any excited state, and are therefore ignored. With this approximation, term C vanishes, leaving only term B in equation (8). Now it is also proved in Albrecht's theory that by combinations and rearrangements of terms in B, one can write this in symmetric form (with respect to e and s) ($E_{ev} - E_{gi} \approx E_{ev} - E_{gj} \approx E_e^0 - E_g^0$ when far away from resonance).

$$(\alpha_{\rho\sigma})_{gi,gj} = B \qquad \text{for } i \neq j \qquad (9)$$

$$B = - \sum_e \sum_{s \neq e} \sum_a \left[[(g^0|R_\rho|e^0)(e^0|h_a|s^0)(s^0|R_\sigma|g^0) \right.$$

$$+ (g^0|R_\sigma|e^0)(e^0|h_a|s^0)(s^0|R_\rho|g^0)]$$

$$\left. \times \frac{[(E_e^0 - E_g^0)(E_s^0 - E_g^0) + E_0^2]\langle i\|Q_a\|j\rangle}{[(E_e^0 - E_g^0)^2 - E_0^2][(E_s^0 - E_g^0)^2 - E_0^2]} \right]$$

The terms in this expression are related to the "forbidden intensity in allowed transitions" in vibronic theory.[21] The principal qualitative result of this theory is the prediction that, as the incident-light frequency approaches a given allowed electronic transition, those normal modes which are vibronically active in the electronic transition should exhibit particularly striking enhancement of their Raman intensities. A test of this prediction would require a combination of a vibronic spectroscopic study (very likely through polarized absorption) with near-resonance Raman studies at wavelengths approaching the electronic transition. No such experimental study has yet been accomplished.

Approximations by Krushinskii and Shorygin. In the semiclassical theory of Shorygin,[9,10] only one excited mode e and one normal mode (a totally symmetric mode) are considered. By setting $\rho = \sigma$, the following expression is obtained in our notation:

$$(\alpha_{\rho\rho})_{gi,gj} \cong A'' + B'' \tag{10}$$

$$A'' = (g^0|R_\rho|e^0)(e^0|R_\rho|g^0)(\partial f/\partial Q)_{Q_0}\langle i\|Q\|j\rangle$$

$$B'' = \{\partial[(g|R_\rho|e)(e|R_\rho|g)]/\partial Q\}_{Q_0}f(Q_0)\langle i\|Q\|j\rangle$$

where

$$f(Q) = \frac{2(E_e - E_g)}{(E_e - E_g)^2 - E_0^2}$$

and

$$\frac{\partial f}{\partial Q} = -\frac{2[(E_e - E_g)^2 + E_0^2]}{[(E_e - E_g)^2 - E_0^2]}\left(\frac{\partial}{\partial Q}(E_e - E_g)\right)$$

In the near-resonance region the factor A'', with the frequency dependence described by $(\partial f/\partial Q)$ in equation (10), appears to be the dominant factor in a number of cases with the exception of some cases where neither A'' nor B'' alone could describe the experimental data.

More recently, Krushinskii and Shorygin presented a "quantum model"[6,7] to modify the semiclassical theory. Assuming $(g|R_\rho|e) = (g^0|R_\rho|e^0)(1 + \eta Q)$, where η is a parameter, they obtained

$$B'' = (g^0|R_\rho|e^0)(e^0|R_\rho|g^0)\left[\frac{2(E_e^0 - E_g^0)}{(E_e^0 - E_g^0)^2 - E_0^2}\right]2\eta\langle i\|Q\|j\rangle \tag{11}$$

And, assuming the potential of the ground state and the excited state to be harmonic and identical in force constant k, but displaced in Q

by the amount Δ, they obtained:[22,23]

$$A'' = (g^0|R_\rho|e^0)(e^0|R_\rho|g^0)\left[\frac{(E_e^0 - E_g^0)^2 + E_0^2}{[(E_e^0 - E_g^0)^2 - E_0^2]^2}\right]2k\Delta\langle i\|Q\|j\rangle \quad (12)$$

Far away from resonance, in the region $(E_e - E_g - E_0) \gg (\omega_0|\Delta|/4|\eta|\langle i\|Q\|j\rangle^2)$,

$$(\alpha_{\rho\rho})_{gi,gj} \cong B''$$

and in the region $(E_e - E_g - E_0) \sim (\omega_0|\Delta|/4|\eta|\langle i\|Q\|j\rangle^2)$,

$$(\alpha_{\rho\rho})_{gi,gj} \cong A'' + B''$$

The main interest in this theory has been in the incident-light frequency dependence of the Raman intensity, fitting experimental data with the equations and deriving the "effective absorption frequencies" to compare with absorption spectra. (See Chapter 6 in Volume 1.)

Savin's Treatment. In 1964 Savin carried out the vibronic expansion in the most general manner without using the two approximations found in the earlier work,[5] and obtained equation (8). Savin further replaced each energy difference $E_{ev} - E_{gi}$ in the denominator by $E_{ev} - E_{gi} = E_{eg} + \Delta E_{ev,gi}$, where E_{eg} is the energy of the purely electronic transitions, and expanded the denominators in a power series of $\Delta E_{ev,gi}$ for the nonresonance case. $\Delta E_{ev,gi}$ is the nuclear-coordinate-dependent factor. By assuming that both the excited-state and the ground-state potentials are harmonic and have the same vibrational frequency, and by also assuming that the potentials are displaced by the amount Δ, it was possible to sum over the vibrational sublevels of each electronic state. In so doing, it was necessary to restrict the theory only to the totally symmetric mode of vibration since nontotally symmetric modes give no displacement of potential without changing the shape of the molecule. Thus, for term A, Savin found

$$A = \sum_{e \neq g}' \left[\frac{(g^0|R_\sigma|e^0)(e^0|R_\rho|g^0)}{(E_{eg} - E_0)^2} + \frac{(g^0|R_\rho|e^0)(e^0|R_\sigma|g^0)}{(E_{eg} + E_0)^2}\right](S') \quad (13)$$

where S', coming from the linear term in $\Delta E_{ev,gi}$, gives intensity to the fundamental lines:

$$S' = -E_v\left(\frac{\mu\omega_v^2\Delta^2}{2E_v}\right)^{\frac{1}{2}}$$

where E_v is the vibrational energy equal to $\hbar\omega_v$ with ω_v being the frequency. The quadratic terms S'' giving intensity to the first overtone line are also quadratic in Δ. We will not be interested in overtones and combination tones in this review. The reader can consult the original paper by Savin for details.

For terms B and C, Savin obtained to first order for the non-resonance case

$$
\begin{aligned}
B = \sideset{}{'}\sum_{e \neq g} \sideset{}{'}\sum_{s \neq e} \sum_a & \left[\frac{(g^0|R_\sigma|e^0)(e^0|h_a|s^0)(s^0|R_\rho|g^0)}{E_{eg} - E_0} \right. \\
& + \frac{(g^0|R_\rho|e^0)(e^0|h_a|s^0)(s^0|R_\sigma|g^0)}{E_{eg} + E_0} + \frac{(g^0|R_\sigma|s^0)(s^0|h_a|e^0)(e^0|R_\rho|g^0)}{E_{eg} - E_0} \\
& \left. + \frac{(g^0|R_\rho|s^0)(s^0|h_a|e^0)(s^0|R_\sigma|g^0)}{E_{eg} + E_0} \right] \frac{\langle i\|Q_a\|j\rangle}{E_e^0 - E_s^0}
\end{aligned}
\tag{14}
$$

$$
\begin{aligned}
C = \sideset{}{'}\sum_{e \neq g} \sideset{}{'}\sum_{t \neq g} \sum_a & \left[\frac{(g^0|h_a|t^0)(t^0|R_\sigma|e^0)(e^0|R_\rho|g^0)}{E_{eg} - E_0} \right. \\
& + \frac{(g^0|h_a|t^0)(t^0|R_\rho|e^0)(e^0|R_\sigma|g^0)}{E_{eg} + E_0} + \frac{(g^0|R_\sigma|e^0)(e^0|R_\rho|t^0)(t^0|R_\sigma|g^0)}{E_{eg} - E_0} \\
& \left. + \frac{(g^0|R_\rho|e^0)(e^0|R_\sigma|t^0)(t^0|h_a|g^0)}{E_{eg} + E_0} \right] \frac{\langle i\|Q_a\|j\rangle}{E_g^0 - E_t^0}
\end{aligned}
$$

which is just what one finds when summing B, C in equation (8) over v if one ignores the vibrational parts of the energy denominators.

A similar treatment was given by Verlan[14,15] in more detail. Verlan expanded the denominators like Savin and kept more terms. The additional terms are shown to be negligible at the nonresonance region. Also discussed in detail were the symmetry properties of the operators and the intermediate states.

Verlan's conclusions can be summed up as follows. Far away from resonance, (1) the totally symmetric mode derives most of its intensity from term A, while the nontotally symmetric mode derives its intensity from only terms B and C; (2) the form of the Raman tensor is practically symmetrical; (3) a change in the symmetry of the molecule in the excited state produces an increase in the intensities of nontotally symmetric vibrations and a deviation in the degree of depolarization.

Discussion of the Vibronic Representations

It is interesting to consider term A in a little more detail here. Suppose we expand the energy denominator around a fixed energy

E_e^0 in a power series:

$$(E_{ev} - E_{gi} - E_0)^{-1} = \sum_{N=0}^{\infty} (-1)^N \left[\frac{(E_{ev} - E_e^0)^N}{(E_e^0 - E_{gi} - E_0)^{N+1}} \right]$$

with

$$\left| \frac{(E_{ev} - E_e^0)}{(E_e^0 - E_{gi} - E_0)} \right| \leq 1$$

Now $(E_{ev} - E_e^0)^N \|v\rangle = (e|(\mathcal{H} - E_e^0)^N|e)\|v\rangle$ because in the adiabatic approximation $(e|\mathcal{H}|e)\|v) = E_{ev}\|v\rangle$. It follows that

$$\sum_v \frac{\langle i\|v\rangle\langle v\|j\rangle}{E_{ev} - E_{gi} - E_0} = \sum_N \sum_v \frac{\langle i\|v\rangle\langle v\|j\rangle(E_{ev} - E_e^0)^N}{(E_e^0 - E_{gi} - E_0)^{N+1}}$$

$$= \sum_N \sum_v \frac{\langle i\|(e|(\mathcal{H} - E_e^0)^N|e)\|v\rangle\langle v\|j)}{(E_e^0 - E_{gi} - E_0)^{N+1}}$$

$$= \sum_N \frac{\langle i\|(e|(\mathcal{H} - E_e^0)^N|e)\|j\rangle}{(E_e^0 - E_{gi} - E_0)^{N+1}} \tag{15}$$

The electronic integral can be expanded in ΔQ. The linear term for $N = 1$ is then

$$\sum_a [\partial(e|\mathcal{H}|e)/\partial Q_a]_{Q_0} \cdot \Delta Q_a$$

or, equivalently,

$$\sum_a [\partial E_e(Q)/\partial Q_a]_{Q_0} \cdot \Delta Q_a \quad \text{or} \quad \sum_a (e^0|h_a|e^0) \cdot \Delta Q_a$$

because, in this case, the order of differentiation, with respect to nuclear coordinates, and integration over electronic coordinates is immaterial.

Upon substituting the linear term of equation (15) into term A, we obtain ($E_e^0 - E_{gi} \approx E_e^0 - E_{gj} \approx E_e^0 - E_g^0$ when far away from resonance):

$$A'' \cong \sum_{e \neq g}' \sum_a \left[\frac{(g^0|R_\sigma|e^0)(e^0|R_\rho|g^0)}{(E_e^0 - E_{gi} - E_0)^2} + \frac{(g^0|R_\rho|e^0)(e^0|R_\sigma|g^0)}{(E_e^0 - E_{gj} + E_0)^2} \right]$$

$$\times (e^0|h_a|e^0)\langle i\|Q_a\|j\rangle$$

$$\cong \sum_{e \neq g} \sum_a \left[\frac{\begin{aligned}(g^0|R_\sigma|e^0)(e^0|h_a|e^0)(e^0|R_\rho|g^0)\\ + (g^0|R_\rho|e^0)(e^0|h_a|e^0)(e^0|R_\sigma|g^0)\end{aligned}}{[(E_e^0 - E_g^0)^2 - E_0^2]^2} \right]$$

$$\times [(E_e^0 - E_g^0)^2 + E_0^2]\langle i\|Q_a\|j\rangle \tag{16}$$

Note that these are precisely the "diagonal" terms $(e = s$ terms) excluded in term B [see equation (9)]. Had Albrecht[5] taken the Q dependence of the energy denominator in term A into account, he might have modified equation (9) by relaxing the restriction $e \neq s$ in the summation, i.e.,

$$(\alpha_{\rho\sigma})_{gi,gj} = -\sum_{e}\sum_{s}\sum_{a}[(g^0|R_\rho|e^0)(e^0|h_a|s^0)(s^0|R_\sigma|g^0)$$

$$+ (g^0|R_\sigma|e^0)(e^0|h_a|s^0)(s^0|R_\rho|g^0)]$$

$$\times \frac{(E_e^0 - E_g^0)(E_s^0 - E_g^0) + E_0^2}{[(E_e^0 - E_g^0)^2 - E_0^2][(E_s^0 - E_g^0)^2 - E_0^2]}\langle i\|Q_a\|j\rangle \quad (9')$$

However, one must realize that term A'' actually is zero for the normal modes that are nontotally symmetric, because $h_a = (\partial\mathscr{H}/\partial Q_a)Q_0$ has the symmetry in electronic space of the normal mode a, and the integral $(e|h_a|e)$ must vanish by symmetry if h_a belongs to a nontotally symmetric representation. In the "quantum model" of Krushinskii and Shorygin, this means that the displacement of the potential Δ in a nontotally symmetric normal coordinate is zero, regardless of the frequency of exciting light. Thus, equation (9) and equation (9′) are identical for the nontotally symmetric modes.

Let us attempt to relate the parameters Δ and η to the vibronic formulation of the problem. The displacement of potential Δ can be related to the matrix element $(e^0|h_a|e^0)$. The potential of the excited state can be given by the following series:

$$V^e(Q) = V^e(Q_0) + \left(\frac{\partial V^e}{\partial Q}\right)_{Q_0}(Q - Q_0) + \frac{1}{2}\left(\frac{\partial^2 V^e}{\partial Q^2}\right)_{Q_0}(Q - Q_0)^2$$

$$= V^e(Q_0) + (e^0|h|e^0)(Q - Q_0) + \frac{1}{2}k_e(Q - Q_0)^2$$

$$= V^e(Q_0) + \frac{1}{2}k_e\left[(Q - Q_0) + \frac{(e^0|h|e^0)}{k_e}\right]^2 - \frac{1}{2}\frac{(e^0|h|e^0)^2}{k_e}$$

One can see that $V^e(Q)$ is a harmonic oscillator potential displaced from Q_0 by $\Delta_e = (e^0|h|e^0)/k_e$. Therefore, $(e^0|h|e^0) = k_e \cdot \Delta_e$. Apparently it is unnecessary to assume that the vibrational frequency is the same for the excited and the ground state. But far away from the resonance region, more than one excited state will have to be considered. For each excited state, there has to be a different displacement of potential Δ_e. Similarly, the parameter η, which also is to be associated with a given excited electronic state, can be identified as follows:

$$\eta_e = \sum_{s \neq e} \left[\frac{(g^0|R_\rho|s^0)(s^0|h_a|e^0)}{(g^0|R_\rho|e^0)(E_e^0 - E_s^0)} \right] = \sum_{t \neq g} \left[\frac{(g^0|h_a|t^0)(t^0|R_\rho|e^0)}{(E_g^0 - E_t^0)(g^0|R_\rho|e^0)} \right] \quad (17)$$

The advantage of the explicit vibronic representation is that the matrix elements in the sum, in certain cases, may be calculated a priori if the excited-state wave functions are available.[24,25]

On the other hand, it is true that term C should also be included in equation (9') for completeness. While the magnitude of term C may not be very large, nevertheless, the states t coupled through $(g^0|h_a|t^0)(t^0|R_\rho|e^0)(e^0|R_\sigma|g^0)$ in term C are distinct from the states s coupled through $(g^0|R_\rho|e^0)(e^0|h_a|s^0)(s^0|R_\sigma|g^0)$ in term B, because t does not have to belong to an allowed transition while s does. For molecules with low-lying forbidden transitions, term C should not be forgotten.

From equation (8), only an allowed transition can give dramatic resonance effect when the incident-light frequency approaches the energy of transition, i.e., when $(E_e - E_g - E_0)$ approaches zero. (Actually, an additional small damping constant $i\gamma_e$ in the denominator must not be neglected as the resonance condition is approached.) While an allowed transition gives a finite transition moment, a forbidden transition gives zero $[(g^0|R_\rho|e^0) = 0]$ in the numerator of all terms in equation (8) and gives no real resonance. This is a generalization of the conclusion drawn by Albrecht from equation (9).

One should expect that this vibronic expansion approach has definite practical value whenever the major Raman intensity is derived from a very small number of excited states (resonance or near-resonance cases), and one must look forward to experimental verification of the expected link between vibronic spectroscopy and Raman intensities in favorable cases. On the other hand, the vibronic representation clearly diverges from the Placzek polarizability theory. The experimental data of ordinary (nonresonance) Raman spectroscopy have unmistakably pointed to the importance of the nature of the chemical bond in the Raman intensities (see Chapter 3, Volume 1). Evidently, there must be great value in the theoretical formulation of Raman intensities as properties of the ground-state wave function. In the second section we return to an examination of the Raman intensity theory within the context of a ground-state approach.

GROUND-STATE APPROACH

Formal Closure of the Sum

Let us look at the infinite sum in equation (4):

$$(\alpha_{\rho\sigma})_{m,n} = \sum_e \left[\frac{\langle m|M_\sigma|e\rangle\langle e|M_\rho|n\rangle}{E_e - E_m - E_0} + \frac{\langle m|M_\rho|e\rangle\langle e|M_\sigma|n\rangle}{E_e - E_n + E_0} \right] \quad (4)$$

This sum is a particular case of a general class of such infinite summations associated with various second-order effects. A recent review of the sum rule for the general summation, and some particular cases, has been given by Hirschfelder *et al.*[26] The Raman sum can be treated in an analogous manner as follows.

Suppose the sums were closed, the closure will transform equation (4) into the following integral:

$$(\alpha_{\rho\sigma})_{m,n} = \langle \varphi^-|M_\rho|n\rangle + \langle m|M_\rho|\varphi^+\rangle \quad (18)$$

It is not hard to see that the functions φ^+ and φ^- have to satisfy the following differential equations:

$$\mathscr{H}|\varphi^-\rangle - (E_m + E_0)|\varphi^-\rangle = M_\sigma|m\rangle$$
$$\mathscr{H}|\varphi^+\rangle - (E_n - E_0)|\varphi^+\rangle = M_\sigma|n\rangle \quad (19)$$

Where $\mathscr{H}$ is the total Hamiltonian with eigenfunctions $|m\rangle, |n\rangle, |e\rangle, \ldots$ for eigenvalues $E_m, E_n, E_e, \ldots$. Let us multiply both sides of equation (19) by $\langle e|$; we get

$$(E_e - E_m - E_0)\langle e|\varphi^-\rangle = \langle e|M_\sigma|m\rangle$$
$$(E_e - E_n + E_0)\langle e|\varphi^+\rangle = \langle e|M_\sigma|n\rangle \quad (20)$$

or (notice that $\langle e|\mathscr{H}|\varphi^\pm\rangle = E_e\langle e|\varphi^\pm\rangle$ because $\mathscr{H}$ is Hermitian)

$$\langle\varphi^-|e\rangle = \frac{\langle m|M_\sigma|e\rangle}{E_e - E_m - E_0}$$
$$\langle e|\varphi^+\rangle = \frac{\langle e|M_\sigma|n\rangle}{E_e - E_n + E_0} \quad (21)$$

Compare equations (4) and (21); we can rewrite equation (4) as

$$(\alpha_{\rho\sigma})_{m,n} = \sum_e [\langle\varphi^-|e\rangle\langle e|M_\sigma|n\rangle + \langle m|M_\rho|e\rangle(e|\varphi^+\rangle)]$$

Now the infinite sum is a simple expression of matrix multiplication and is easily closed to give equation (18):

$$(\alpha_{\rho\sigma})_{m,n} = \langle \varphi^- | M_\sigma | n \rangle + \langle m | M_\rho | \varphi^+ \rangle$$

Difficulties arise only in solving the differential equation (19) for the functions $|\varphi^-\rangle$ and $|\varphi^+\rangle$. These differential equations are at least as difficult to solve as the Schroedinger equation, and are, as expected, not yet solved exactly for any molecule.

Transformation of the Sum

As explained above, the infinite sum in equation (4) is very difficult to solve exactly because of the differential equation (19). However, there is a way to avoid the differential equations completely.[18] Let us expand the denominators of equation (4) in a power series around some average energy E_{av} as above:

$$\frac{1}{E_e - E_m - E_0} = \sum_N (-1)^N \frac{(E_e - E_{av})^N}{(E_{av} - E_m - E_0)^{N+1}}$$

$$\frac{1}{E_e - E_n + E_0} = \sum_N (-1)^N \frac{(E_e - E_{av})^N}{(E_{av} - E_n + E_0)^{N+1}}$$

with

$$\left| \frac{(E_e - E_{av})}{(E_{av} - E_m(E_n) - E_0)} \right| < 1$$

$$(\alpha_{\rho\sigma})_{m,n} = \sum_e \sum_N (-1)^N \left[\frac{\langle m | M_\sigma | e \rangle \langle e | M_\rho | n \rangle (E_e - E_{av})^N}{(E_{av} - E_m - E_0)^{N+1}} \right.$$

$$\left. + \frac{\langle m | M_\rho | e \rangle \langle e | M_\sigma | n \rangle (E_e - E_{av})^N}{(E_{av} - E_n + E_0)^{N+1}} \right] \qquad (22)$$

Now, E_e is the eigenvalue of the total Hamiltonian $\mathcal{H}$ for the eigenfunction $|e\rangle$, i.e., $\mathcal{H}|e\rangle = E_e|e\rangle$, and, therefore, $(\mathcal{H} - E_{av})^N|e\rangle = (E_e - E_{av})^N|e\rangle$ so that $\langle m|M_\sigma|e\rangle(E_e - E_{av})^N = \langle m|M_\sigma(\mathcal{H} - E_{av})^N|e\rangle$. Thus,

$$(\alpha_{\rho\sigma})_{m,n} = \sum_e \sum_N (-1)^N \left[\frac{\langle m | M_\sigma (\mathcal{H} - E_{av})^N | e \rangle \langle e | M_\rho | n \rangle}{(E_{av} - E_m - E_0)^{N+1}} \right.$$

$$\left. + \frac{\langle m | M_\rho (\mathcal{H} - E_{av})^N | e \rangle \langle e | M_\sigma | n \rangle}{(E_{av} - E_n + E_0)^{N+1}} \right] \qquad (23)$$

Exchange the order of the summations and note that the denominators are now constant in the summation of e. Equation (23) becomes

$$(\alpha_{\rho\sigma})_{m,n} = \sum_N (-1)^N \left\{ \frac{\sum_e \langle m|M_\sigma(\mathcal{H} - E_{av})^N|e\rangle\langle e|M_\rho|n\rangle}{(E_{av} - E_m - E_0)^{N+1}} \right.$$
$$\left. + \frac{\sum_e \langle m|M_\rho(\mathcal{H} - E_{av})^N|e\rangle\langle e|M_\sigma|n\rangle}{(E_{av} - E_n + E_0)^{N+1}} \right\} \qquad (24)$$

The summations of e in the numerators are now only matrix multiplications and are easily closed, giving

$$(\alpha_{\rho\sigma})_{m,n} = \sum_N (-1)^N \left\{ \frac{\langle m|M_\sigma(\mathcal{H} - E_{av})^N M_\rho|n\rangle}{(E_{av} - E_m - E_0)^{N+1}} \right.$$
$$\left. + \frac{\langle m|M_\rho(\mathcal{H} - E_{av})^N M_\sigma|n\rangle}{(E_{av} - E_n + E_0)^{N+1}} \right\} \qquad (25)$$

Thus, we have eliminated the sums over e without having to solve the differential equations (equation 19). Instead, the infinite sum over N remains to be evaluated. This sum will not even converge if for any E_e the inequality necessary for the geometric expansion [see equation (22) above] is violated. What is to be emphasized, however, is that the introduction of E_{av}, together with the geometric expansion, offers a parameter with which one can, in principle, isolate the major contributions to $(\alpha_{\rho\sigma})_{m,n}$ in just a few terms. Thus, since $(\alpha_{\rho\sigma})_{m,n}$ itself cannot depend on E_{av}, neither can the complete expansion of equation (25). A variation of E_{av} serves only to control the distribution of the contributions to the polarizability among the terms in equation (25). The hope is that a suitable choice of E_{av} will serve to concentrate the major contribution into the leading terms. A theoretical evaluation of these terms with good ground-state wave functions would then constitute a good theoretical basis for the prediction of Raman intensities. In contrast, equation (4) would call for knowledge of the complete set of molecular wave functions and, instead, encourage reference to experimental vibronic spectroscopy, as we have seen.

Adiabatic Approximation

In vibrational Raman spectroscopy, it is assumed that the adiabatic approximation is applicable for the initial and the final states, $|m\rangle$ and $|n\rangle$. In other words, it is meaningful to write the wave functions of $|m\rangle$

and $|n\rangle$ as products of electronic and vibrational parts; thus [as in equation (5a) above],

$$|m\rangle = |g\rangle\||i\rangle \quad \text{or} \quad \psi_m(r, Q) = \theta_g(r, Q)\phi_i(Q)$$

$$|n\rangle = |g\rangle\||j\rangle \quad \text{or} \quad \psi_n(r, Q) = \theta_g(r, Q)\phi_j(Q) \tag{26}$$

Substituting into equation (25), we obtain

$$
\begin{aligned}
(\alpha_{\rho\sigma})_{gi,gj} &= \sum_N (-1)^N \left[\frac{\langle i\|(g|M_\sigma(\mathscr{H} - E_{av})^N M_\rho|g)\||i\rangle}{(E_{av} - E_{gi} - E_0)^{N+1}} \right. \\
&\quad \left. + \frac{\langle i\|(g|M_\rho(\mathscr{H} - E_{av})^N M_\sigma|g)\||j\rangle}{(E_{av} - E_{gj} + E_0)^{N+1}} \right] \\
&= \langle i\| \sum_N (-1)^N \left[\frac{(g|M_\sigma(\mathscr{H} - E_{av})^N M_\rho|g)}{(E_{av} - E_{gi} - E_0)^{N+1}} \right. \\
&\quad \left. + \frac{(g|M_\rho(\mathscr{H} - E_{av})^N M_\sigma|g)}{(E_{av} - E_{gj} + E_0)^{N+1}} \right] \||j\rangle
\end{aligned}
$$

or

$$(\alpha_{\rho\sigma})_{gi,gj} = \langle i\|\alpha_{\rho\sigma}(Q)\||j\rangle \tag{27}$$

We thus have defined the ground-state polarizability tensor $\alpha_{\rho\sigma}(Q)$ as a function of nuclear coordinates

$$
\begin{aligned}
\alpha_{\rho\sigma}(Q) &= \sum_N (-1)^N \left[\frac{(g|M_\sigma(\mathscr{H} - E_{av})^N M_\rho|g)}{(E_{av} - E_{gi} - E_0)^{N+1}} \right. \\
&\quad \left. + \frac{(g|M_\rho(\mathscr{H} - E_{av})^N M_\sigma|g)}{(E_{av} - E_{gj} + E_0)^{N+1}} \right]
\end{aligned} \tag{28}
$$

When Q is equal to Q_0, the equilibrium nuclear coordinates in the ground electronic state, $\alpha_{\rho\sigma}(Q_0)$ is precisely the expectation value of the polarizability. When $\alpha_{\rho\sigma}(Q)$ is expanded with respect to a normal coordinate of vibration Q_a, we obtain as the first-order coefficient the polarizability derivative $(\partial\alpha_{\rho\sigma}/\partial Q_a)_{Q_0}$, which, for a fundamental line, gives the following relationship:

$$(\alpha_{\rho\sigma})_{g0,g1} = \langle 0\|\alpha_{\rho\sigma}(Q)\||1\rangle = (\partial\alpha_{\rho\sigma}/\partial Q_a)_{Q_0}\langle 0\|\Delta Q_a\||1\rangle$$

All the "ground-state theories" discussed here aim at the calculation of $(\partial\alpha_{\rho\sigma}/\partial Q_a)_{Q_0}$.

It is interesting to note that if we define the average energy by

$$E_{av} = \frac{(g|M_\sigma \mathcal{H} M_\rho|g)_0}{(g|M_\sigma M_\rho|g)_0} \tag{29}$$

where the subscript denotes that the integral is evaluated at $Q = Q_0$, we find that the $N = 1$ term vanishes, and if we neglect all the higher terms, $N \geq 2$, we obtain the following approximations:

$$\alpha_{\rho\sigma}(Q_0) = \frac{(g|M_\sigma M_\rho|g)_0}{E_{av} - E_{gi} - E_0} + \frac{(g|M_\rho M_\sigma|g)_0}{E_{av} - E_{gj} + E_0}$$

$$= \frac{(g|M_\sigma M_\rho|g)_0^2}{(g|M_\sigma(\mathcal{H} - E_{gi} - E_0)M_\rho|g)_0}$$

$$+ \frac{(g|M_\rho M_\sigma|g)_0^2}{(g|M_\rho(\mathcal{H} - E_{gj} + E_0)M_\sigma|g)_0} \tag{30}$$

which turns out to be identical to the expressions used in a number of calculations for polarizability. However, in these previous cases, the expression was obtained by the variation method.[23]

Variation Method.[27-32] To demonstrate the variation method for polarizability calculations let us go back to equation (4) and let $|m\rangle = |n\rangle$ and $\rho = \sigma$; then we have

$$(\alpha_{\sigma\sigma})_{m,m} = \sum_e \left[\frac{\langle m|M_\sigma|e\rangle\langle e|M_\sigma|m\rangle}{E_e - E_m - E_0} + \frac{\langle m|M_\sigma|e\rangle\langle e|M_\sigma|m\rangle}{E_e - E_m + E_0} \right]$$

$$= \langle \varphi^-|M_\sigma|m\rangle + \langle m|M_\sigma|\varphi^+\rangle \tag{31}$$

The corresponding differential equations for φ^- and φ^+ are

$$\mathcal{H}|\varphi^-\rangle - (E_m + E_0)|\varphi^-\rangle = M_\sigma|m\rangle$$
$$\mathcal{H}|\varphi^+\rangle - (E_m - E_0)|\varphi^+\rangle = M_\sigma|m\rangle \tag{32}$$

Let the trial variation function be in the following forms (C_- and C_+ are the variation parameters):

$$|\varphi^-\rangle = C_- M_\sigma|m\rangle$$
$$|\varphi^+\rangle = C_+ M_\sigma|m\rangle \tag{33}$$

In a symmetric form, one can rewrite the sums[28,29] as $(\alpha_{\sigma\sigma})_{mm} \cong \alpha^- + \alpha^+$, and

$$
\begin{aligned}
\alpha^- &= \langle\varphi^-|M_\sigma|m\rangle + \langle m|M_\sigma|\varphi^-\rangle - \langle\varphi^-|\mathcal{H} - E_m - E_0|\varphi^-\rangle \\
\alpha^+ &= \langle\varphi^+|M_\sigma|m\rangle + \langle m|M_\sigma|\varphi^+\rangle - \langle\varphi^+|\mathcal{H} - E_m + E_0|\varphi^+\rangle
\end{aligned}
\tag{34}
$$

Since $\alpha_{\sigma\sigma} = \alpha^- + \alpha^+$, being the polarizability, is proportional to the energy, one can apply the variation principle to minimize α^- and α^+ by varying the parameters C_- and C_+. Substitute equation (33) into equation (34), differentiate α^- and α^+ with respect to C_- and C_+, and set the derivative to zero. One then finds

$$
\left(\frac{\partial\alpha^\pm}{\partial C_\pm}\right) = 2\langle m|M_\sigma M_\sigma|m\rangle - 2C_\pm\langle m|M_\sigma(\mathcal{H} - E_m \pm E_0)M_\sigma|m\rangle = 0
$$

or

$$
C_\pm = \frac{\langle m|M_\sigma M_\sigma|m\rangle}{\langle m|M_\sigma(\mathcal{H} - E_m \pm E_0)M_\sigma|m\rangle}
\tag{35}
$$

so that

$$
(\alpha_{\sigma\sigma})_{m,m} \cong \left[\frac{\langle m|M_\sigma M_\sigma|m\rangle^2}{\langle m|M_\sigma(\mathcal{H} - E_m - E_0)M_\sigma|m\rangle}\right.
$$
$$
\left. + \frac{\langle m|M_\sigma M_\sigma|m\rangle^2}{\langle m|M_\sigma(\mathcal{H} - E_n + E_0)M_\sigma|m\rangle}\right]
\tag{36}
$$

Comparing equations (36) and (30), it is evident that equation (36) is a special case of equation (30).

In 1935 Hirschfelder calculated the polarizability of the H_2 molecule[29] at different bond lengths using the variation method and then interpolated for the derivative of the polarizability in nuclear coordinates. Similarly, in 1950 Bell and Long repeated the same calculations with different wave functions and included the H_2^+ ion in the calculation.[30] Of course, the variation function can be much more elaborate than the one suggested by equation (33), and one finds very elaborate calculations for the H_2 molecule in the literature.[31,32] However, the variation wave function given by equation (33) is the simplest, and applications to other molecules for the calculations of Raman intensities are more or less related to this approximation, i.e., equations (30) and (36).

Delta-Function Model. The delta-function-potential model has been used in a series of studies[32,37] of the properties of chemical binding. Recently, Lippincott and Stutman[38] extended the model to the calculations of polarizabilities of various diatomic and polyatomic molecules. Subsequently, Long and Plane[16] went on to compute the derivatives of polarizabilities for the symmetrical stretching modes of molecules and ions. Similar calculations were also published by Lippincott and Nagarajan[17] with a different method for obtaining the so-called "delta-function-strength" parameters. The results of both sets of calculations are not identical.

The basic approximation of the delta-function-potential model is the replacement of the Coulomb potential of each nucleus by a normalized delta function times a parameter which is determined by the effective charge of the nucleus. In the calculations of Long and Plane, this parameter appears as g/Z, where Z is the magnitude of the unshielded nuclear charge and g is the "delta-function-potential strength," taken to be the square root of the valence-state electronegativity on the Pauling scale, $g = \chi^{\frac{1}{2}}$. Moreover, the square root of this parameter, $(g/Z)^{\frac{1}{2}}$, is supposed to be equal to the charge seen by an electron. For heteronuclear diatomic systems, $(g/Z)^{\frac{1}{2}}$ is taken to be the geometric mean of the two atomic values.

In the calculations of Lippincott and Nagarajan, however, this same parameter appears as A, the "one-electron delta-function strength" for an atom. The value of A for an atom is related to the separated atom energies, which, in turn, are related to the electronegativities. However, in practice A is obtained from an empirical rule as follows[37]:

$$A = [\chi/(2.6n - 1.7p - 0.8D + 3.0F)]^{\frac{1}{2}} \qquad (37)$$

where χ is the electronegativity of the atom, n is the principal quantum number, $p = 1$ for atoms with p electrons in the valence shell and 0 for atoms with no p electrons in the valence shell, D is the total number of completed p and d shells in an atom, and F is the total number of completed f shells in an atom. In the case of heteronuclear diatomic systems, the root-mean-square delta-function strength of the two nuclei A_{12} is used.

In both calculations, equation (36) was used to obtain an expression for polarizability when $E_0 = 0$ (static polarizability). A direct differentiation of the expression with respect to bond length gives the polarizability derivative. It is assumed that only the component of the polarizability that is parallel to the bond $\partial\alpha_\parallel/\partial R$ contributes to the polarizability derivative, and, therefore, the averaged polarizability derivative $(\partial\bar{\alpha}/\partial R)$ is assumed to be one third of the value of the parallel component. The two final expressions are given as follows:

$$(\partial\bar{\alpha}/\partial R) = \tfrac{1}{3}n(g/Z)(\sigma/a_0)(R^3) \quad \text{(Long and Plane)} \quad (38a)$$

$$(\partial\bar{\alpha}/\partial R) = \tfrac{1}{3}NA_{12}(\sigma/a_0)(R^3) \quad \text{(Lippincott and Nagarajan)} \quad (38b)$$

where $n = 2N$ = number of bonding electrons, a_0 is the Bohr radius, $\sigma = \exp\left[-(\chi_1 - \chi_2)^2/4\right]$ is the Pauling covalent-bond character, and R is the bond length. Results of both calculations are tabulated to-gether in Table 1. The two sets of calculated values exhibit only

Table 1. Delta-Function Calculations of the Polarizability Deriva-tive for Several Molecules and Ions [Equation (36) as a Basis], Symmetric Stretch Vibrations ($\partial\bar{\alpha}/\partial R$ in Å^2)

Molecule	Calculated*	Experiment*	Experiment†	Calculated†
CH_4	1.20	1.03	1.04	0.684
$C_2H_6(C-C)$	1.81	0.92 or 1.37	1.37	1.729
(C-H)	1.20	1.08 or 1.10	1.08	0.715
$C_2H_4(C-C)$	2.46	1.89	1.89	2.244
(C-H)	1.11	1.04	1.04	0.647
$C_2H_2(C-C)$	2.94	3.36	2.92	3.275
(C-H)	0.87	1.05	1.02	0.634
CCl_4	2.02	2.04	2.08	2.435
$CHCl_3(C-Cl)$	1.90	1.92	1.93	2.465
$CH_2Cl_2(C-Cl)$	1.75	1.73	1.73	2.460
$SiCl_4$	2.06	1.96	1.96	2.303
$GeCl_4$	2.28	2.66	2.66	2.474
$SnCl_4$	3.08	3.37	3.37	3.184
CBr_4	2.76	3.33	3.33	3.024
$SnBr_4$	4.02	6.73	6.73	3.893
Ions				
$CO_3^=$	1.05	1.04	1.08	1.1371
NO_3^-	0.83	1.64	1.71	1.466
$PO_4^=$	0.85	0.91	0.95	1.122
ClO_3^-		1.39	1.39	1.558
BrO_3^-		1.95	1.95	2.459
IO_3^-		2.24	2.24	2.231

*From T. V. Long and R. A. Plane.[16]
†From E. R. Lippincott and G. Nagarajan.[17]

moderate agreement. In some cases the differences are rather large (e.g., CH_4, NO_3^-). But in view of the serious approximations and the empirical nature of some of the parameters, it is rather remarkable that most of the values do give a certain degree of agreement with the experimental values. Moreover, this delta-function-potential model is the first theoretical model to give some relationship between the intensity of a Raman line and the nature of the chemical bond (degree of covalent character), in agreement with earlier observations (see Chapter 4, Volume 1).

Direct Differentiation. Let us return to equation (28). We have here the complete expression for the ground-state polarizability as a function of nuclear coordinates:

$$\alpha_{\rho\sigma}(Q) = \sum_N (-1)^N \left[\frac{(g|M_\sigma(\mathcal{H} - E_{av})^N M_\rho|g)}{(E_{av} - E_{gi} - E_0)^{N+1}} \right.$$
$$\left. + \frac{(g|M_\rho(\mathcal{H} - E_{av})^N M_\sigma|g)}{(E_{av} - E_{gj} + E_0)^{N+1}} \right]$$

One can differentiate $\alpha_{\rho\sigma}(Q)$ term by term to obtain the complete polarizability derivative. Thus,

$$\left(\frac{\partial\alpha_{\rho\sigma}}{\partial Q_a}\right)_{Q_0} = \sum_N (-1)^N \left[\frac{\{(\partial/\partial Q_a)(g|M_\sigma(\mathcal{H} - E_{av})^N M_\rho|g)\}_{Q_0}}{(E_{av} - E_{gi}^0 - E_0)^{N+1}} \right.$$
$$\left. + \frac{\{(\partial/\partial Q_a)(g|M_\rho(\mathcal{H} - E_{av})^N M_\sigma|g)\}_{Q_0}}{(E_{av} - E_{gj}^0 + E_0)^{N+1}} \right] \qquad (39)$$

Recalling the success of the calculations by an equivalent method of keeping only the $N \leq 1$ term of equation (39) and setting $E_{av} = (g|M_\sigma\mathcal{H}M_\rho|g)_0/(g|M_\sigma M_\rho|g)_0$ in the variation approaches, we now will study equation (39) directly with terms $N < 2$, assuming that the terms $N \geq 2$ will lead to a small remainder generally. For simplicity of comparison, the absolute intensities are discussed in terms of the polarizability derivative $(\partial\alpha_{\rho\sigma}/\partial Q_a)_0$ extrapolated to zero frequency of exciting light, i.e., $E_0 = 0$. The small difference of energies E_{gi} and E_{gj} are also ignored in the following expression taken to order $N < 2$:

$$\left(\frac{\partial\alpha_{\rho\sigma}}{\partial Q_a}\right)_{Q_0} = 2\frac{\{(\partial/\partial Q_a)(g|M_\sigma M_\rho|g)\}_{Q_0}}{E_{av} - E_g^0}$$
$$- 2\frac{\{(\partial/\partial Q_a)(g|M_\sigma(\mathcal{H} - E_{av})M_\rho|g)\}_{Q_0}}{(E_{av} - E_g^0)^2} \qquad (40)$$

The function $|g\rangle$, as stated earlier in the derivation, is the ground-state electronic wave function under the adiabatic approximation. In other words, $|g\rangle$ is the eigenfunction to the electronic Hamiltonian H_{el}, which formally is just the total Hamiltonian $\mathscr{H}$ with the kinetic energy of the nuclei T_N missing. In equation (40), the integral

$$(g|M_\sigma(\mathscr{H} - E_{av})M_\rho|g) = (g|M_\sigma(H_{el} + T_N - E_{av})M_\rho|g)$$

If we add and subtract a term $E_g(Q)(g|M_\sigma M_\rho|g)$, we can rewrite the integral as

$$(g|M_\sigma(\mathscr{H} - E_{av})M_\rho|g) = (g|M_\sigma(H_{el} - E_g)M_\rho|g)$$
$$- (E_{av} - E_g)(g|M_\sigma M_\rho|g) + (g|M_\sigma T_N M_\rho|g) \qquad (41)$$

It can be shown that the derivative of the first term $(g|M_\sigma(H_{el} - E_g)M_\rho|g)$ vanishes if $E_g(Q)$ is the eigenvalue of the operator $H_{el}(r, Q)$ operating on the eigenfunction $|g\rangle$ (see Appendix B). Also, by the definition of Q_0, $(\partial E_g(Q)/\partial Q)_{Q_0}$ is zero. Moreover, the last term, $(g|M_\sigma T_N M_\rho|g)$, is analogous to the correction to the adiabatic approximation and can be shown to be negligible compared with the first term.[18] Thus, we obtain the following simple equation, correct to order $N < 2$:

$$\left(\frac{\partial \alpha_{\rho\sigma}}{\partial Q_a}\right)_{Q_0} = \frac{4}{E_{av} - E_g^0}\left\{\frac{\partial}{\partial Q_a}(g|M_\sigma M_\rho|g)\right\}_{Q_0} \qquad (42)$$

or, by definition of E_{av},

$$\left(\frac{\partial \alpha_{\rho\sigma}}{\partial Q_a}\right)_{Q_0} = \frac{4(g|M_\sigma M_\rho|g)_0}{(g|M_\sigma(\mathscr{H} - E_g)M_\rho|g)_0}\left\{\frac{\partial}{\partial Q_a}(g|M_\sigma M_\rho|g)\right\}_{Q_0} \qquad (43)$$

Now since we know that this choice of E_{av} gives a good approximation to $\alpha_{\rho\sigma}^0$, i.e.,

$$\alpha_{\rho\sigma}^0 \cong \frac{2(g|M_\sigma M_\rho|g)_0}{E_{av} - E_g^0} \qquad (44)$$

we can also write

$$\left(\frac{\partial \alpha_{\rho\sigma}}{\partial Q_a}\right)_{Q_0} \cong 2\alpha_{\rho\sigma}^0 \frac{((\partial/\partial Q_a)(g|M_\sigma M_\rho|g))_{Q_0}}{(g|M_\sigma M_\rho|g)_0} \qquad (45)$$

where $\alpha_{\rho\sigma}^0$ can be obtained from measurements or calculations in the literature. When $\alpha_{\rho\sigma}^0$ is not readily available, one can also make use of the following identity (see Appendix B) for any ρ,σ correct to the order of the adiabatic approximation:

$$(g|M_\sigma(H_{el} - E_g)M_\rho|g)_0 = \frac{n}{2}(e^4 a_0) \tag{46}$$

where n is the number of electrons in the Hamiltonian and obtain with equation (43).

$$\left(\frac{\partial\alpha_{\rho\sigma}}{\partial Q_a}\right)_{Q_0} = \frac{8}{n}(e^4 a_0)^{-1}(g|M_\sigma M_\rho|g)^0 \left(\frac{\partial}{\partial Q_a}(g|M_\sigma M_\rho|g)\right)_{Q_0} \tag{47}$$

Equation (47) is identical to equation (43) only when the wave function $|g)$ is an eigenfunction of H_{el}. In general, the true eigenfunctions are not available, and calculations based on equations (43) and (47) will give different results for any approximate wave function. Equations (43), (45), and (47) offer particularly simple methods for calculating intensities.

Calculations for H_2^+ ion and H_2 using different wave functions are tabulated in Tables 2 and 3, together with the results of the calculations by other methods. In the case of the H_2 molecule, the experimental value is also included. It is interesting to note that the James function for H_2^+ ion gives very close results for equations (43) and (47), reflecting the fact that this function also gives excellent energy calculations, while the poorer LCAO-MO functions for H_2^+ and H_2 give larger discrepancies.

On closer examination of the calculations of the H_2 molecule, a more drastic approximation suggests itself.[18] The familiar atomic orbital based molecular wave function Ψ, rather naturally appears

Table 2. Polarizability Derivative Calculations[18] for H_2^+

Method	$\alpha_x', \text{Å}^2$	$\alpha_z', \text{Å}^2$	$E_g, e^2/a_0$*
James function			
Bell and Long[30]	0.31	1.90	
equation (43)	0.31	1.94	−0.6021
equation (47)	0.31	1.95	
LCAO-MO function			
Bell and Long[30]	0.34	2.02	
equation (43)	0.36	2.58	−0.583
equation (47)	0.38	3.45	

*The exact ground-state energy is −0.6027.

Table 3. Polarizability Derivative Calculations[18] for H_2

Source	$\bar{\alpha}^0$, 10^{-24} cm^3	$(\partial\alpha/\partial R)_0$, Å^2	ρ_{Ram} (depol.)	Description of work
Bell and Long[30]*	0.72	1.49	0.018	LCAO-MO, 1 parameter
Present work	0.72	1.456	0.100	LCAO-MO, equation (43)
	0.72	1.837	0.088	LCAO-MO, equation (45), experimental $\alpha^0_{\rho\sigma}$ used
	0.72	1.909	0·212	LCAO-MO, equation (47)
Hirschfelder[29]	0.68	0.93	0.012	Wang, 1 parameter, V.B.
	0.69	0.89	0.003	Rosen, 1 parameter, polarized AO
	0.73	0.99	0.003	Wang, 2 parameter
	0.74	1.00	0.002	Rosen, 2 parameter
Ishiguro et al.[32]†	0.741	1.411	0.045	James–Collidge, 11 + 10 and 11 + 9 terms, rotations averaged
	(0.7894)	($\alpha_{01} = 0.139$)	(0.052)	
Kolos et al.[31]†	0.8023	(1.146)	(0.082)	54 + 34 terms, rotations averaged
	(0.8045)			
Experiment I[39]	0.8056	1.13	0.070	Extrapolated to $E_0 = 0$
Experiment II[40]	—	1.30	0.073	Not extrapolated

*Bond length taken to be 1.375 a_0.
†Values in brackets are averaged over rotations in room temperature.

as a product of two parts, the unnormalized function of atomic orbitals Ψ and the normalization factor N, which is an explicit function of atomic overlap. It appears that the major contribution to the polarizability derivative is from the overlap integrals in the normalization factor rather than from the details of the wave function itself. In other words,

$$\left\{\frac{\partial}{\partial Q}(\Psi|M_\sigma M_\rho|\Psi)\right\}_{Q_0} \cong \left(\frac{\partial N^2}{\partial Q}\right)_{Q_0} (\psi|M_\sigma M_\rho|\psi)_0 \qquad (48)$$

The results of calculations for H_2, based on equation (48) with a series of simple functions, are given in Table 4. The results are, in general, surprisingly good compared to the calculations in Table 3, considering the serious approximations involved. Even more remarkable is the agreement between the diverse types of atomic-orbital-based functions used, (VB and MO), which seems to imply that the major contribution to the derivative is indeed from the normalization factor of the wave function, rather than from the details of the wave function.

Table 4. Approximate Calculations[18] for H_2

Simple functions (see Tang and Albrecht[18] for details)	$\bar{\alpha}^0$, $10^{-24}\,cm^3$	$N^{-1}(\partial N/\partial R)_0$, a_0^{-1}	$(\partial\bar{\alpha}/\partial R)_0$, Å^2		ρ_{Ram} (depol.)	
			I*	II†	I*	II†
a) VB	1.305	0.1327	1.309	0.808	0.001	0.020
b) MO	1.599	0.1576	1.905	0.960	0.042	0.020
c) Wang	0.717	0.1524	0.826	0.928	0.001	0.020
d) Coulson	0.881	0.1998	1.331	1.217	0.079	0.020
e) Weinbaum I	0.737	0.1832	1.021	1.116	0.018	0.020
f) Weinbaum II	0.723	0.1698	0.927	1.034	0.010	0.020
Experiment I‡	0.8056	—	1.13		0.070	
Experiment II**	—	—	1.30		0.073	

*Calculated from equation (47).
†Calculated from equation (45) using experimental value for $\alpha_{\rho\sigma}^0$.
‡From Yoshino and Bernstein[39]; values are extrapolated to zero frequency of exciting light.
**Golden and Crawford[40]; not extrapolated.

Since the functions used are quite representative of typical covalent-bond wave functions, the same approximation, equation (48), may prove useful in molecules other than H_2.

Discussion of the Ground-State Approach

A discussion of the ground-state representation as related to experimental observation has been given[18] along the following lines.

A basic question in Raman spectroscopy is the relationship between the intensity of a given Raman line and the nature of the chemical bonds which participate in the corresponding normal mode. A number of empirical rules have been developed in this connection and we shall comment on some of these on the basis of the results from the direct differentiation of the polarizability expression.

Within the framework of the bond polarizability theory[41–43] (discussed in Volume 1), it has been established for some time that the mean polarizability derivative of a bond A–B, $\bar{\alpha}'_{AB}$, depends on the bond type—it is small for a purely ionic bond and much larger for a purely covalent bond. Woodward and Long[44] found that, for the totally symmetric modes of the tetrahalides of Group-IV elements, $\bar{\alpha}'_{AB}$ is proportional to the percent covalent character of the bond and the sum of the atomic number of the two bonded atoms. The work of

Yoshino and Bernstein[45] with hydrocarbon gases revealed that $\bar{\alpha}'_{AB}$ is directly proportional to bond order. The delta-function-potential model was able to lead to a direct proportionality of polarizability derivative to bond order and the degree of covalent character. However, other studies of inorganic ions indicate that σ and π bonds do not always contribute equally to $\bar{\alpha}'_{AB}$.[46]

Another empirical rule of practical importance in analyzing the Raman spectrum is that the totally symmetric vibrations generally exhibit higher intensities than the nontotally symmetric vibrations. At least one attempt to rationalize this rule has appeared.[47] Although this rule seems to be quite successful, in general, and has been one of the more powerful tools in analyzing Raman spectra, one cannot always rely on it since a few exceptions are known.

Many of the empirical rules imply additivity properties of the polarizability derivative. This in turn implies some kind of factorization of the ground-state wave function. In Appendix C a factorization is given such that the n-electron ground-state function gives

$$|g) = (n!)^{-\frac{1}{2}} A \psi(1, 2, \ldots, n)$$

where $\psi(1, 2, \ldots, n) = \prod_i \varphi_i(i_1, i_2, \ldots, i_m)$ and A is the antisymmetrizer. The many-electron functions φ_i are functions for "noninteracting" fragments of the molecule. These fragments conceivably can range from inner-core electrons of atoms to independent ions, and from localized bonds to highly delocalized functions. It is shown[18] (Appendix C) how neglect of appropriate terms leads to additivity of polarizability and, through equation (42), to $\alpha'_{\rho\sigma} = \sum_i (\alpha'_{\rho\sigma})_i$, giving additivity of the polarizability derivative. This presumably must be the theoretical starting point for understanding the various empirical additivity rules (bond polarizability theory, specific contributions from bond types, etc.). Appendix C also shows how the *relative motion* of suitably partitioned fragments *cannot give rise to any polarizability derivative*. The only source of Raman intensity must come from the relative motion of the nuclei within the fragments themselves. Thus, the motion of monatomic fragments (such as atomic core fragments) cannot contribute to the Raman intensity. The relative motion of ions (to the extent that they can appear as separate factors in the "primitive" wave function ψ) cannot give rise to significant Raman intensity. This simple observation is very likely a satisfactory explanation of why so-called highly ionic bonds exhibit weak Raman intensity. *The internal relative motion of nuclei* in a polyatomic fragment (ion or otherwise) must provide *the major contribution to Raman intensity*. The covalent bond is a primary example of such a polyatomic (diatomic) fragment, while the purely ionic bond, in zeroth order, is not.

Let us now examine the source of Raman intensity through a somewhat different approach, especially in keeping with the recent work. We note that any normalized electronic wave function may be written as a product of the normalization factor $N(Q)$, and an un-normalized function $\psi(Q, r)$. Thus, with $\Psi = N\psi$, it is possible to write the derivative of the polarizability $\alpha' = \alpha'_N + \alpha'_\psi$. This partitioning of α' is, of course, quite arbitrary since N may be absorbed into the definition of ψ to any extent one pleases. However, as found in the studies presented in the last section, a certain "natural" factoring appears when atomic-orbital-based functions are used and, at least for the six H_2 functions examined, the factoring seems to give useful results. This leads us to hope that, for the frequently used atomic-orbital-based functions, an isolation of the normalization factor from the wave function may be possible such that equation (48) is applicable. Whenever one can factor a wave function into a Q-dependent part only and a primitive function ψ, one can show that α'_N *can be nonzero only for totally symmetric vibrations*. And to the extent that one can succeed in writing $\Psi = N(Q)\psi(r, Q)$ such that $\alpha'_N \gg \alpha'_\psi$ one then has found a rationale for the fact that totally symmetric vibrations generally exhibit greater Raman intensity than nontotally symmetric modes. Thus,

$$\alpha' \text{ (totally symmetric)} = \alpha'_N + \alpha'_\psi$$

$$\alpha' \text{ (nontotally symmetric)} = \alpha'_\psi$$

and if $\alpha'_N \gg \alpha'_\psi$ in general, as seems to be possible, α' (totally symmetric) $\gg \alpha'$ (nontotally symmetric). The symmetry argument follows upon first noting that for $\partial(\Psi|M_\sigma M_\rho|\Psi)/\partial Q_a$ to be nonvanishing, $M_\sigma M_\rho$ must be a basis for the same representation of the group for which Q_a is a basis in nuclear-coordinate space. (Incidentally, when necessary in these arguments, $\rho\sigma$ can be regarded as a component of a symmetry transform from the Cartesian tensor.) Now the matrix element in electron space for α'_N is just[18] $(\psi|M_\sigma M_\rho|\psi)_0$; for nontotally symmetric vibrations $M_\sigma M_\rho$ is nontotally symmetric and this integral must vanish. Therefore, while α'_N may be the significant term in each bond of the polyatomic molecule, the contributions must cancel for the nontotally symmetric case when the whole molecule is taken into consideration. Hopefully, studies of polyatomic molecules along these lines will serve to evaluate this particular explanation for the relatively high intensity of the totally symmetric modes in a spectrum. In view of this discussion and the results of the calculations presented here it seems very worthwhile to extend such calculations to simple polyatomic molecules where more than one normal mode will appear, including nontotally symmetric vibrations.

APPENDIX A

The All-Particle Dispersion Equation for Molecules in the Transition Moment Representation. The Adiabatic Approximation and Neglect of Infrared Virtual States

It is a straight-forward matter that the matrix elements of the momentum operator for the kth particle (mass μ_k, charge q_k) in a system, $\mathbf{P}_k$, can be written in terms of the dipole moment operator $\mathbf{R}_k$ (charge times position vector of kth particle). For the σth component of this vector and for the kth particle, one has for the meth matrix element

$$\langle m|(P_k)_\sigma|e\rangle = \frac{\mu_k}{q_k} \cdot \frac{i}{\hbar} \langle m|\mathscr{H}(R_k)_\sigma - (R_k)_\sigma\mathscr{H}|e\rangle \qquad \text{(A-1)}$$

or

$$\langle m|(P_k)_\sigma|e\rangle = \frac{\mu_k}{q_k} \cdot \frac{i}{\hbar}(E_m - E_e)\langle m|(R_k)_\sigma|e\rangle \qquad \text{(A-2)}$$

since $\langle m|$ and $|e\rangle$ are eigenstates of $\mathscr{H}$. The charge-to-mass ratio in equation (A-2) can be incorporated into the left-hand side and a sum taken over all particles in the system to give

$$\langle m|\sum_k \frac{q_k}{\mu_k}(P_k)_\sigma|e\rangle = \frac{i}{\hbar}(E_m - E_e)\langle m|R_\sigma|e\rangle \qquad \text{(A-3)}$$

where $\sum_k (R_k)_\sigma \equiv R_\sigma$.

Now equation (2) of the text, properly generalized to *all* particles, should have the charge-to-mass ratio absorbed into each matrix element of the numerator. The summation is over all particles, electrons as well as nuclei. The typical such factor in the numerator would then look like the left-hand side of equation (A-3) above. Making this substitution we obtain the general *all-particle* expression in terms of transition dipoles.

$$(\alpha_{\rho\sigma})_{m,n} = \sum_e{}' \left[\frac{\langle m|R_\sigma|e\rangle\langle e|R_\rho|n\rangle}{E_e - E_m - E_0} + \frac{\langle m|R_\rho|e\rangle\langle e|R_\sigma|n\rangle}{E_e - E_n + E_0} \right] \quad \text{(A-4)}$$

The introduction of equation (A-3) into the generalized equation (2) is carried out stepwise, and proper algebraic treatment with closure of the sum over states e shows how ω and ω_0 in equation (2) are exactly eliminated. The prime in the summation equation (A-4) means that states m and n must be excluded in the sum over e. It is seen how the substitution given by equation (A-3) vanishes for a diagonal matrix element. Since the algebraic manipulation leading to equation (A-4) has lost this zero, the exclusion on the sum must be explicit. And this is necessary only if state m or n has a dipole moment. Equation (A-4) is just equation (3) of the text.

Up to this point the eigenstates of $\mathcal{H}$ are genuine eigenstates in nuclear-electron space and the eigenvalues will form a nearly continuous spectrum. At this level of treatment there is no formal distinction between electronic and vibrational levels, and the Raman spectroscopy is molecular in the true sense of the word. Furthermore, because of the nearly continuous nature of the eigenstates, most Raman spectroscopy is near or on resonance, strictly speaking. A damping factor in the denominator should not be eliminated. We know, in fact, that the adiabatic approximation works quite well and that there are spectral regions where the transition moments into vibronic levels are negligibly strong and nonresonance Raman scattering is a reality. On introducing the adiabatic approximation into equation (A-4) it is useful at the same time to recognize that R_σ contains R_σ', the dipole operators for electrons only, and R_σ'', those for the heavy particles, the charged nuclei. Let us then make the following replacements in equation (A-4).

$$R_\sigma = R_\sigma' + R_\sigma''$$

$$|m\rangle = |g\rangle\|i^g\rangle$$

$$|n\rangle = |g\rangle\|j^g\rangle$$

$$|e\rangle = |e\rangle\|v^e\rangle$$

The sum over e (vibronic) in equation (A-4) is now a sum over all *electronic* states e, and all vibrational sublevels of each, excluding only sublevels i and j of the ground electronic state—for we wish to focus upon vibrational Raman spectroscopy of the ground electronic state. Alternate choices for i and j could lead to vibrational Raman spectroscopy of excited electronic states or to electronic Raman spectroscopy. The sum over electronic states is conveniently split into two parts: $g \neq e$ and $g = e$. The former reads simply

$$(\alpha_{\rho\sigma})_{i^g,j^g}^{g \neq e} = \sum_{e \neq g, v^e} \frac{\langle i^g \| (g|R'_\sigma|e) \| v^e \rangle \langle v^e \| (e|R'_\rho|g) \| j^g \rangle}{E_{e,v^e} - E_{g,i^g} - E_0} + (\text{second term})$$

(A-5)

The nuclear dipole-moment operator R''_σ plays no role in these $e \neq g$ terms because of orthogonality in electron space. Equation (A-5) is therefore an *electrons-only* part of the dispersion equation in the adiabatic approximation and invariably has been the basis for dealing with ordinary Raman scattering. Let us briefly examine the $g = e$ part of the sum. This term reads

$$(\alpha_{\rho\sigma})_{i^g,j^g}^{g = e} = \sum_{v^g \neq i^g, j^g} \frac{\langle i^g \| (g|R'_\sigma + R''_\sigma|g) \| v^g \rangle \langle v^g \| (g|R'_\rho + R''_\rho|g) \| j^g \rangle}{E_{v^g} - E_{i^g} - E_0}$$

$$+ (\text{second term})$$

(A-6)

Because of the exclusion $v^g \neq i^g$, j^g on the sum, v^g must be simultaneously dipole coupled in lowest order to *both* i^g and j^g. If i^g and j^g are different states then this must mean that these two states are separated by *two* vibrational quanta in the harmonic approximation. Thus, the infrared virtual states contribute only to combination and overtone Raman spectroscopy and equation (A-6) can be neglected, as it is, in the scattering of fundamentals.

Order-of-magnitude estimates of the various terms are of interest. In the scattering of fundamentals (equation A-5) one of the electronic transition moments appearing in the numerator must be taken to first order in nuclear displacement. Such a first-order term represents a reduction in dipole strength of the order of the fourth root of the electron to nuclear mass ratio χ. (The vibronic perturbation energy is roughly an electronic energy $\overline{e^2/r}$ times $\overline{Q/r}$ or a nuclear displacement to electron-displacement ratio which goes as $\chi^{\frac{1}{4}}$.) The contribution from equation (A-6) amounts to a reduction of dipole strength (over pure electronic dipole strength) of second order or $\chi^{\frac{1}{2}}$ (this is clear either directly for the R''_σ, R''_ρ terms or through necessary vibronic expansions of the *both* $(g|R'_\sigma|g)$ terms). Thus, with scattered intensity proportional to the square of the polarizability, we can estimate that the relative strength of *electronic Raman scattering : fundamental vibrational Raman scattering : combination and overtone vibrational Raman scattering* goes as $1 : \chi^{\frac{1}{2}} : \chi$—or roughly as $1 : 10^{-2} : 10^{-4}$. The question of the energy denominators can considerably alter this picture, certainly as any resonance condition is approached.

APPENDIX B

To show

$$\left\{\frac{\partial}{\partial Q_a}(g|M_\sigma(H_{el} - E_g)M_\rho|g)\right\}_{Q_0} = 0$$

or in simpler notation,

$$\left\{\frac{\partial}{\partial Q_a}F_{\rho\sigma}(Q)\right\}_{Q_0} = F'_{\rho\sigma}(Q_0) = 0 \tag{B-1}$$

we have the following: Given that

$$H_{el}|g) = E_g|g)$$

$$H_{el} = \sum_{j=1}^{n} \frac{P_j^2}{2\mu_e} + V(q, Q) \tag{B-2}$$

for n electrons. Then

$$(H_{el} - E_g)M_\rho|g) = (H_{el}M_\rho - M_\rho H_{el})|g) \tag{B-3}$$

M_ρ commutes with $V(q, Q)$, giving

$$(H_{el}M_\rho - M_\rho H_{el})|g) = \sum_j \frac{1}{2\mu_e}(P_j^2 M_\rho - M_\rho P_j^2)|g)$$

$$= \sum_j \frac{1}{2\mu_e}(P_j(P_j M_\rho) + P_j M_\rho P_j - M_\rho P_j^2)|g)$$

$$= \sum_j \frac{1}{2\mu_e}(P_j(P_j M_\rho) + (P_j M_\rho)P_j)|g) \tag{B-4}$$

so that

$$F_{\rho\sigma} = \sum_j \frac{1}{2\mu_e}(g|M_\sigma P_j(P_j M_\rho) + M_\sigma(P_j M_\rho)P_j|g) \tag{B-5}$$

Now,

$$P_j = \frac{\hbar}{i} \nabla_j, \qquad P_j^* = -P_j$$

and P_j is Hermitian; therefore,

$$(g|M_\sigma P_j(P_j M_\rho)|g) = (g|(P_j M_\rho)P_j^* M_\sigma|g)$$
$$= -(g|(P_j M_\rho)(P_j M_\sigma)|g) - (g|(P_j M_\rho)M_\sigma P_j|g)$$

to give

$$F_{\rho\sigma}(Q) = -\sum_j \frac{1}{2\mu_e}(g|(P_j M_\rho)(P_j M_\sigma)|g) \qquad (B\text{-}6)$$

Now M_σ and M_ρ are defined by

$$M_\sigma = \sum_k -er_{\sigma k} - (g|\sum_l -er_{\sigma l}|g), \qquad M_\rho = \sum_k -er_{\rho k} - (g|\sum_l -er_{\rho l}|g)$$

where the electronic integrals are functions of nuclear coordinates only, so that

$$-\sum_j \frac{1}{2\mu_e}(P_j M_{\sigma k})(P_j M_{\rho l}) = \sum_{j=1}^n \frac{e^2\hbar^2}{2\mu_e}\delta_{jk}\delta_{jl} = \frac{n\hbar^2 e^2}{2\mu_e} \qquad (B\text{-}7)$$

Equation (B-6) now becomes

$$F_{\rho\sigma}(Q) = \frac{n\hbar^2 e^2}{2\mu_e}(g|g) = \frac{n}{2}(e^4 a_0)(g|g) \qquad (B\text{-}8)$$

Since the factor $(n/2)(e^4 a_0)$ is a constant, the derivative of $F_{\rho\sigma}(Q)$ is

$$F'_{\rho\sigma}(Q_0) = \frac{n}{2}(e^4 a_0)\left\{\frac{\partial}{\partial Q}(g|g)\right\}_{Q_0} \qquad (B\text{-}9)$$

But since

$$\left\{\frac{\partial}{\partial Q}(g|g)\right\}_{Q_0} = 2(g^0|g')$$

and

$$|g'\rangle = \sum_{e^0}' \frac{(e^0|(\partial H_{el}/\partial Q)_{Q_0}|g^0)}{E_e^0 - E_g^0}|e^0)$$

then

$$(g^0|g') = \sum_{e^0}{}' \frac{(e^0|(\partial H_{el}/\partial Q)_{Q_0}|g^0)}{E_e^0 - E_g^0}(g^0|e^0) = 0$$

thus proving equation (B-1).

APPENDIX C

Additivity of Polarizability

In general, an n-electron wave function is written as follows:

$$|g) = (n!)^{-\frac{1}{2}}A\Psi(1, 2, \ldots, n)$$

where the operator A is the antisymmetrizer which permutes the n electrons. Recalling the definition of the operators $R' = -er$ (and dropping the prime for convenience),

$$M_\sigma = \sum_{k=1}^{n} R_{\sigma k} - (g| \sum_{l=1}^{n} R_{\sigma l}|g)$$

$$M_\rho = \sum_{k=1}^{n} R_{\rho k} - (g| \sum_{l=1}^{n} R_{\rho l}|g) \tag{C-2}$$

(the sum is over all electrons), in a perfectly general manner, we can write the n-electron integral $(g|M_\sigma M_\rho|g)$ as follows:

$$(g|M_\sigma M_\rho|g) = (\Psi|M_\sigma M_\rho|A\Psi)$$

$$= (\Psi| \sum_{k} (R_\sigma R_\rho)_k|A\Psi) + (\Psi| \sum_{k} \sum_{l \neq k} R_{\sigma k} R_{\rho l}|A\Psi)$$

$$-(\Psi| \sum_{k} R_{\sigma k}|A\Psi)(\Psi| \sum_{l} R_{\rho l}|A\Psi) \tag{C-3}$$

Ψ is the "primitive" function which we assume separable into non-interacting groups,

$$\Psi(1, 2, \ldots, n) = \prod_{i} \varphi_i(i_1, i_2, \ldots, i_m) \tag{C-4}$$

The ith group, containing electrons i_1 to i_m, and the jth group, containing electrons j_1 to j_m, say, are assumed to occupy different regions of space. Substituting this function into equation (C-3), we obtain

$$(g|M_\sigma M_\rho|g) = \sum_i \left[(\varphi_i| \sum_{k=i_1}^{i_m} (R_\sigma R_\rho)_k| A\varphi_i) + (\varphi_i| \sum_{\substack{k=i_1 \\ k \neq l}}^{i_m} \sum_{l=i_1}^{i_m} R_{\sigma k} R_{\rho l}| A\varphi_i) \right]$$

$$+ \sum_i \sum_j (\varphi_i \varphi_j| \sum_{k=i_1}^{i_m} \sum_{l=j_1}^{j_m} (R_{\sigma k} R_{\rho l} + R_{\rho k} R_{\sigma l})| A\varphi_i \varphi_j)$$

$$- \sum_i \sum_j (\varphi_i| \sum_{k=i_1}^{i_m} R_{\sigma k}| A\varphi_i)(\varphi_j| \sum_{l=j_1}^{j_m} R_{\rho l}| A\varphi_j) \qquad (C-5)$$

Notice that in the third term the operator is written such that when R_σ is operating on an electron in φ_i, R_ρ is operating on an electron in φ_j. Terms exclusive of electron permutation between the partitioned fragments give integrals which cancel the cross terms ($i \neq j$) in the last sum, while terms involving electron permutations between the partitioned fragments vanish since integration over k (or l) involves the overlap of φ_i and φ_j which we have taken to be negligible. Thus, we obtain

$$(g|M_\sigma M_\rho|g) = \sum_i \left\{ (\varphi_i| \sum_{k=i_1}^{i_m} (R_\sigma R_\rho)_k + \sum_{\substack{k=i_1 \\ k \neq l}}^{i_m} \sum_{l=i_1}^{i_m} R_{\sigma k} R_{\rho l}| A\varphi_i) \right.$$

$$\left. - (\varphi_i| \sum_{k=i_1}^{i_m} R_{\sigma k}| A\varphi_i)(\varphi_i| \sum_{i=i_1}^{i_m} R_{\rho l}| A\varphi_i) \right\} \qquad (C-6)$$

It follows from equation (42), for derived polarizability,

$$\left(\frac{\partial \alpha_{\rho\sigma}}{\partial Q} \right)_{Q_0} = \frac{4}{E_{av} - E_g} \left\{ \frac{\partial}{\partial Q}(g|M_\sigma M_\rho|g) \right\}_{Q_0} = \sum_i (\alpha'_{\rho\sigma})_i \qquad (C-7)$$

Now it can easily be demonstrated for all i that term i in equation (C-6) is invariant with respect to an arbitrary displacement of fragment i. Thus the sole contribution to $(\alpha'_{\rho\sigma})_i$ can come from the *relative* motion of nuclei within the ith separated fragment. In other words, the relative motion of the partitioned fragments cannot contribute to Raman intensity. Thus, any partitioned fragment identifiable with only one nucleus can give no Raman intensity whatsoever. The best example of such a fragment is a core function or a monatomic ion. It follows that the relative motion of "true ions" cannot generate Raman intensity any more than can the relative motion of any of the partitioned fragments. Reviews of the bases for factorization of the wave function can be found in Daudel[48] and Primas.[49] Conditions for optimizing a factorization can be raised. In this respect it is interesting to note that the condition for a factorization, such that additivity of polarizability is successful, is the minimization of integrals of the type $(\varphi_i(1)\varphi_j(2)|R_{\sigma 1}R_{\rho 2}|\varphi_i(2)\varphi_j(1))$ which we have taken to be zero here.

REFERENCES

1. H. A. Kramers and W. Heisenberg, *Z. Physik* **31**: 681 (1925).
2. P. A. M. Dirac, *Proc. Roy. Soc. (London)* **114**: 710 (1927).
3. J. H. Van Vleck, *Proc. Natl. Acad. Sci. U.S.* **15**: 754 (1929).
4. G. Placzek, in: E. Marx, *Handbuch der Radiologie*, Akademische Verlagsgesellschaft, Leipzig (1934), Vol. 6, p. 205; English translation by Ann Werbin, UCRL Trans-526 (L), is processed by the clearing house for Federal Scientific and Technical Information of the U.S. Department of Commerce.
5. A. C. Albrecht, *J. Chem. Phys.* **34**: 1476 (1961).
6. L. L. Krushinskii and P. P. Shorygin, *Opt. i Spektroskopiya* **11**: 24 (1961) [*Opt. Spectry.* **11**: 12 (1961)].
7. L. L. Krushinskii and P. P. Shorygin, *Opt. i Spektroskopiya* **11**: 151 (1961) [*Opt. Spectry.* **11**: 80 (1961)].
8. L. L. Krushinskii and P. P. Shorygin, *Opt. i Spektroskopiya* **19**: 562 (1965) [*Opt. Spectry.* **19**: 312 (1965)].
9. P. P. Shorygin, *Dokl. Akad. Nauk SSSR* **87**: 201 (1952).
10. P. P. Shorygin, *Izv. Akad. Nauk SSSR Ser. Fiz.* **17**: 581 (1953).
11. F. A. Savin, *Opt. i Spektroskopiya* **19**: 555 (1965) [*Opt. Spectry.* **19**: 308 (1965)].
12. F. A. Savin, *Opt. i Spektroskopiya* **19**: 743 (1965) [*Opt. Spectry.* **19**: 412 (1965)].
13. F. A. Savin, *Opt. i Spektroskopiya* **20**: 989 (1966) [*Opt. Spectry.* **20**: 549 (1966)].
14. E. M. Verlan, *Opt. i Spektroskopiya* **20**: 605 (1966) [*Opt. Spectry.* **20**: 341 (1966)].
15. E. M. Verlan, *Opt. i Spektroskopiya* **20**: 802 (1966) [*Opt. Spectry.* **20**: 447 (1966)].
16. T. V. Long and R. A. Plane, *J. Chem. Phys.* **43**: 457 (1965).
17. E. R. Lippincott and G. Nagarajan, *Bull. Soc. Chim. Belges.* **74**: 551 (1965).
18. J. Tang and A. C. Albrecht, *J. Chem. Phys.* **49**: 1144 (1968).
19. G. Herzberg and E. Teller, *Z. Physik. Chem. (Leipzig)* **B21**: 558 (1933).
20. I. I. Kondilenko, P. A. Korotkov, and V. L. Strizhevskii, *Opt. i Spektroskopiya* **9**: 26 (1960) [*Opt. Spectry.* **9**: 13 (1960)].
21. A. C. Albrecht, *J. Chem. Phys.* **33**: 156 (1960).
22. K. K. Rebane and T. K. Rebane, *Izv. Akad. Nauk Est. SSR, Ser. Fiz.-Mat. i Tekh. Nauk* **12**: 227 (1963).
23. R. A. Preem, *Tr. Inst. Fiz. i Astron., Akad. Nauk Est. SSR* **20**: 114 (1963); **25**: 47 (1964).
24. A. D. Liehr, *Z. Naturforsch.* **13a**: 596 (1958).
25. A. C. Albrecht, *J. Chem. Phys.* **33**: 169 (1960).
26. J. O. Hirschfelder, W. Byers Brown, and S. T. Epstein, *Advances in Quantum Chemistry* (1964), Vol. I, p. 256, see also, A. Dalgarno, *Rev. Mod. Phys.* **35**: 522 (1963).
27. M. Karplus, *J. Chem. Phys.* **41**: 880 (1964); Kolker and Karplus **39**: 2011 (1963).
28. M. Karplus and H. J. Kolker, *J. Chem. Phys.* **39**: 1493 (1963); *ibid.* **39**: 2997 (1963).
29. J. O. Hirschfelder, *J. Chem. Phys.* **3**: 555 (1935).
30. R. P. Bell and D. A. Long, *Proc. Roy Soc. (London)* **A203**: 364 (1950).
31. W. Kolos and L. Wolniewiez, *J. Chem. Phys.* **46**: 1426 (1967).

32. E. Ishiguro, T. Arai, M. Mizushima, and M. Kotani, *Proc. Roy. Soc.* (*London*) **A65**: 178 (1952).
33. K. Ruedenberg and R. G. Parr, *J. Chem. Phys.* **19**: 1268 (1951); *ibid.* **21**: 1565 (1953).
34. A. A. Frost, *J. Chem. Phys.* **25**: 1150 (1956).
35. A. A. Frost and F. E. Leland, *J. Chem. Phys.* **25**: 1154 (1956).
36. E. R. Lippincott, *J. Chem. Phys.* **23**: 603 (1955); *ibid.* **26**: 1678 (1957).
37. E. R. Lippincott and M. O. Dayhoff, *Spectrochim. Acta* **16**: 807 (1960).
38. E. R. Lippincott and J. M. Stutman, *J. Chem. Phys.* **68**: 2926 (1964).
39. T. Yoshino and H. J. Bernstein, *J. Mol. Spectry.* **2**: 213 (1958).
40. D. M. Golden and B. Crawford, *J. Chem. Phys.* **36**: 1654 (1962).
41. D. A. Long, *Proc. Roy. Soc.* (*London*) **A217**: 203 (1953).
42. M. V. Vol'kenstein, M. A. El'yashevich, and B. I. Stepanov, *Molecular Vibrations*, State Publishers of Technical-Theoretical Literature, Moscow-Leningrad (1949), Vol. II.
43. M. V. Vol'kenstein, *Compt. Rend. Acad. Sci. URSS* **32**: 185 (1941).
44. L. A. Woodward and D. A. Long, *Trans. Faraday Soc.* **45**: 1131 (1949).
45. T. Yoshino and H. J. Bernstein, *Petroleum Conference on Molecular Spectroscopy*, Pergamon Press, New York (1960).
46. G. W. Chantry and R. A. Plane, *J. Chem. Phys.* **32**: 319 (1960).
47. I. I. Kondilenko and V. L. Strizhevskii, *Opt. i Spektroskopiya* **17**: 528 (1964) [*Opt. Spectry.* **17**: 285 (1964)].
48. R. Daudel, *Advances in Quantum Chemistry* (1964), Vol. I, p. 115.
49. H. Primas, in O. Sinanoglu, *Modern Quantum Chemistry*, Academic Press, New York, Vol. 2, p. 45.

Chapter 3

Raman Spectroscopy with Laser Excitation

H. W. Schrötter

Sektion Physik der Universität München
Germany

INTRODUCTION

In the first volume of this book Koningstein[1] treated the subject of laser Raman spectroscopy. In his article he described the development of the method to about the point where it became apparent that the use of lasers for the excitation of Raman spectra is not only equivalent to mercury lamp excitation, but far superior. The results which were obtained since then have fully justified this hope. They are so numerous that it is not possible to give a complete account of all published material in a relatively short article. Many results on the light scattering in crystals, including the electronic Raman effect, can be found in a conference report.[2]

This chapter is restricted to the linear Raman effect. The stimulated and other nonlinear Raman effects which are excited with high-power pulsed lasers have been treated in review articles[3,4] and in a book[5] recently. The latter also contains a review of the linear Raman effect with laser excitation, which has been the subject of an article by Brandmüller,[6] and progress reports.[7] The instruments available until the summer of 1967 have been described by Schrader.[8] An account of many new developments is contained in a book on Raman spectra by Sushchinskii.[103]

In the first part of this chapter the present state of experimental techniques will be described, including a short description of commercial instruments. In the second and third parts a selection of results will be presented, obtained from liquid, gaseous, and solid samples.

EXPERIMENTAL TECHNIQUES

A modern Raman spectrometer consists of a laser, a sample compartment, a double monochromator, and a photoelectric detection system. Some important aspects of these parts will be treated in the following, concluding with a review of complete Raman spectrometers.

Lasers

The He–Ne laser is the most frequently used light source, with a power ranging from 3 to 180 mW at 632.8 nm. The main advantage of this type of laser is—besides its reliability—its wavelength, which allows the investigation of many colored substances without interference from absorption or fluorescence.

The most powerful source is the argon ion laser. Its strong emission lines at 488.0 and 514.5 nm fall in a region of high sensitivity of the photomultipliers and allow completely new experiments in Raman spectroscopy. The technological difficulties of the early models have been overcome by the electrodeless ring discharge or by graphite, metal-segmented, or ceramic discharge tubes. Commercial models have a power of the order of 1 W in one of the lines mentioned above. Further lines with smaller power are at 457.9, 476.5, 496.5, and 501.7 nm, and can be used for studies of the dependence of Raman intensities on the exciting wavelength.

The krypton ion laser (exciting lines at 647.1, 568.2, 530.8, 520.8, 482.5, and 476.2 nm) may one day replace the He–Ne laser due to its greater versatility. Also, the xenon ion laser has been used as a Raman source.

Mooradian and Wright[9] were the first to use a continuous neodymium laser with a power of 1 W at 1064.8 nm for the excitation of the Raman spectra of plasmons in GaAs. Since their original work the power has been increased to the order of 100 W.[95]

The quasicontinuous ruby laser[10] is becoming a very useful Raman source. The average power of up to 1 W offsets the lower sensitivity of most photomultipliers at the exciting wavelength of 694.3 nm compared with He Ne excitation at 632.8 nm. The longer wavelength, however, is useful for the investigation of substances which absorb too strongly at 632.8 nm.[11] The pulse frequency of 50 cps allows the use of a linear gate which is only opened during the laser emission and suppresses the noise of the photomultiplier in the remaining time; such an electronic system is described below. Delhaye[12] is able to record a complete Raman spectrum from every single pulse of this laser, and to study chemical reactions in this way.

The cadmium laser, the first commercially available metal-vapor laser, emits 50 mW at 441.6 nm and up to 15 mW at 325.0 nm and may be used where the high power of the argon laser is not required, and for cw excitation in the UV, where previously only the pulsed-nitrogen laser (337.1 nm) was available.

Sample Illumination and Collection of Scattered Radiation

Sample compartments and sample cells should be constructed in such a way that the best possible use is made of the exciting laser light and the scattered Raman light from small sample volumes. According to calculations by Delhaye and Migeon[13] and by Benedek and Fritsch[14] this is achieved by focussing the laser beam into the sample with a lens of short focal length f (e.g., a microscope objective) and by projecting an enlarged image of this focus on the slit of the spectrometer. A thorough treatment of the problem has been given by Barrett and Adams[15] and recently by Schwiesow.[127]

The bright region of the focus of a single mode (TEM$_{00}$) laser beam with diameter d has the form of a cylinder ("source cylinder") with radius

$$w_0 = \frac{2\lambda f}{\pi d}$$

and length

$$2b = \frac{16\lambda f^2}{\pi d^2}$$

where λ is the wavelength (see Fig. 1).

To select the focal length of the focussing lens the following considerations are necessary. Usually a spectrometer with a solid

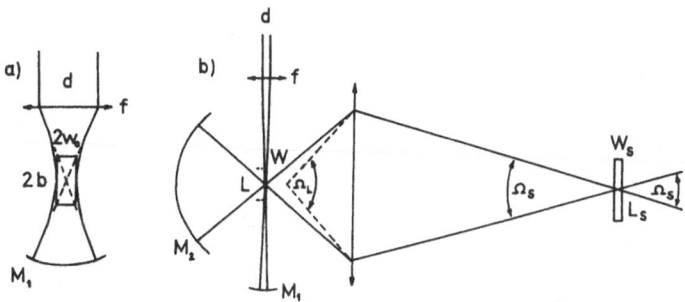

Fig. 1. a) Sample illumination and b) collection of scattered radiation.

angle Ω_s subtended by its pupil is given. Then one has to find a collecting lens with the largest possible solid angle of collection Ω_L. The magnification M of the source image on the slit is then limited by the relation

$$(M + 1)^2 \leq \Omega_L/\Omega_s$$

To obtain the desired resolution in the Raman spectrum the slit width W_s must be chosen. The width of the source cylinder is then determined by

$$2w_0 \leq W = W_s/M$$

In order to focus the beam to a source cylinder of diameter W one has to use a focussing lens with

$$f = \frac{d\pi W}{4\lambda} = \frac{d\pi W_s}{4\lambda M}$$

In this derivation it is assumed that the slit height L_s can be adjusted to match the length L of the source cylinder:

$$L = L_s/M = 2b$$

The total useful Raman flux collected is then proportional to

$$R = \frac{W_s^2 \Omega_s}{4\lambda} = W_s \Omega_s \left(\frac{ML_s}{16\pi\lambda}\right)^{\frac{1}{2}}$$

By a suitable concave mirror M_1 the laser beam can be reflected in itself and the flux of the exciting light nearly doubled. Another factor of almost two in Raman flux can be gained by mirror M_2 which collects the light scattered backward. A field lens in front of the slit is helpful.

However, in practice a limit is set to the observable Raman flux by the "thermal lens effect."[16] Already very slight absorption of the sample leads to a temperature increase in the focus, a corresponding decrease of refractive index, and the formation of a thermal lens which defocusses the laser beam.

There is one disadvantage in the focussing of the laser beam and in the big solid angle of the collecting lens. As Bridge and Buckingham[17] have shown, depolarization ratios of nearly completely polarized lines are measured with a considerable systematic error, when the laser beam is focussed by a lens with $f \lesssim 100$ mm and the light scattered into too big a solid angle is collected. This difficulty will be discussed later in greater detail.

Double Monochromators

It is now generally recognized that only a double monochromator will give good and complete Raman spectra of all kinds of samples. Most commercial double monochromators are part of complete Raman spectrometers and will be described in a later section.

In Germany two double monochromators are available commercially which are suitable for Raman spectroscopy. RSV in Hechendorf, a subsidiary of Carl Zeiss in Oberkochen, offers a 1-m double Czerny–Turner with straight or curved horizontal slits of 8 cm length. The gratings are mounted on a common axis. A linear wave-number drive is being developed. Steinheil–Lear Siegler AG, Munich, has constructed a new series of grating spectrographs and monochromators with lens optics. The double monochromator Spectrovar DM 700 has a focal length of 0.8 m and a light gathering power of the collimator ($f = 0.45$ m) of 1:4.1. The two gratings are arranged on a common pivot table; standard slit height is 25 mm. A linear wave-number drive with speeds from 3.4 to 5400 cm^{-1}/min is available.

Photoelectric Detection Systems

The comparatively weak Raman signals are detected by photomultipliers, amplified, and recorded. For the blue–green region of the spectrum, e.g., argon laser excitation, many types of photomultipliers with high quantum efficiency and low dark current are available. Photomultipliers which are sensitive for red light (S-20, the newly developed S-25, and S-1 response characteristics), however, have a relatively high dark current. Its fluctuations lead to a high noise level in the recorded spectra. In the following the methods to improve the signal-to-noise ratio shall be discussed.

Topp and co-workers[18] derive a formula for the signal-to-noise ratio:

$$\frac{S}{N} = \left[\frac{c\pi RCN_S}{a^2 \left(1 + \dfrac{N_B}{N_S} + \dfrac{N_D^*}{N_S} \right)} \right]^{\frac{1}{2}}$$

where c is the efficiency of the electrostatic focussing of the cathode current to the first dynode, RC the time constant of the detection system, N_S the count rate of photoelectrons arising from the signal, N_B the count rate of photoelectrons arising from the optical background,

$$a^2 = 1 + \frac{1}{g_1} + \frac{1}{g_1(g - 1)}$$

after Sharpe,[19] with g_1 the gain of the first dynode stage and g the average gain of the remaining dynode stages, and

$$N_D^* = N_D + \frac{I_{Dy}}{eG} + \frac{\overline{I_{RL}^2}}{2a^2 e^2 G^2} + \frac{I_L}{eG}$$

N_D^* contains the different sources of dark-current noise: the rate N_D of actual dark-current electrons emitted from the photocathode, the noise electrons arising in the dynode chain (I_{Dy} is the dark current from the dynodes, e is the charge of the electron, and G is the overall gain of the dynode chain), the thermal noise of the load resistance R_L ($\overline{I_{RL}^2} = 4kT\Delta f/R_L$; Δf is the bandpass), and the leakage current I_L at the photomultiplier socket.

To improve the signal-to-noise ratio, the numerator of the equation for S/N must be increased and the denominator decreased. N_B is very effectively suppressed by the use of a double monochromator.[20] Since a^2 essentially decreases with increasing gain of the first dynode stage, it is advisable to keep the voltage between photocathode and first dynode constant at its optimum value with a zener diode. This also results in an optimum value of c.

Because a stems from the statistical fluctuations of the emission of secondary electrons at the dynodes, it is equal to 1 when a pulse counting system is used. Instead of recording the photocurrent as in dc detection systems, in photon counting the number of photoelectrons is recorded. By pulse shaping the contribution of every photoelectron to the recorded signal is made equal.

An additional advantage of the photon counting is the possibility of suppressing the contribution of noise electrons from the dynodes to N_D^* by a discriminator. The noise electrons from the dynodes give rise to smaller pulses than electrons from the cathode, so that a discriminator voltage can be selected which allows only the higher pulses from the cathode to be amplified.

The third term in the expression for N_D^* is always very small and can be neglected. The fourth term may become as important as the first, especially when the photomultiplier socket becomes moist by cooling. When the load resistance R_L is small, as in the case of pulse counting, the fluctuations of I_L become small and can also be neglected.

So we are left with N_D as the most important noise source. According to Richardson's law the dark current is strongly temperature dependent and proportional to the area of the photocathode. Therefore the effective photocathode area should be kept small, either by use of a multiplier with small effective cathode (EMI 6256 S with S-11 response for blue sensitivity or ITT "star-tracker" types) or by magnetic defocussing. The signal light is concentrated on the center of the photocathode by a lens or by a light-pipe cone.[21] The dark-current electrons from the outer parts of the photocathode are deviated from the first dynode by a magnetic field from a solenoid placed around the photocathode[21] or by a permanent ring magnet in front of it.[22,8]

The use of a light-pipe cone allows us, in addition, to make use of multiple reflections at the photocathode.[23] Figure 2 shows the system employed by Topp and co-workers.[18] The light-pipe cone made of lucite is surrounded by a cylinder from the same material. The cone is silvered, but total reflection could also be used. The aluminized face of the cylinder is optically contacted to the front of the photomultiplier by an immersion oil. A permanent ring magnet gives even better results than the coil alone.

Figure 3 shows part of the Raman spectrum of crystalline azo-benzene which demonstrates the improvement in signal-to-noise ratio through this arrangement using dc detection and the Raman spectrometer described by Brandmüller and coworkers.[24] Spectrum *a* was

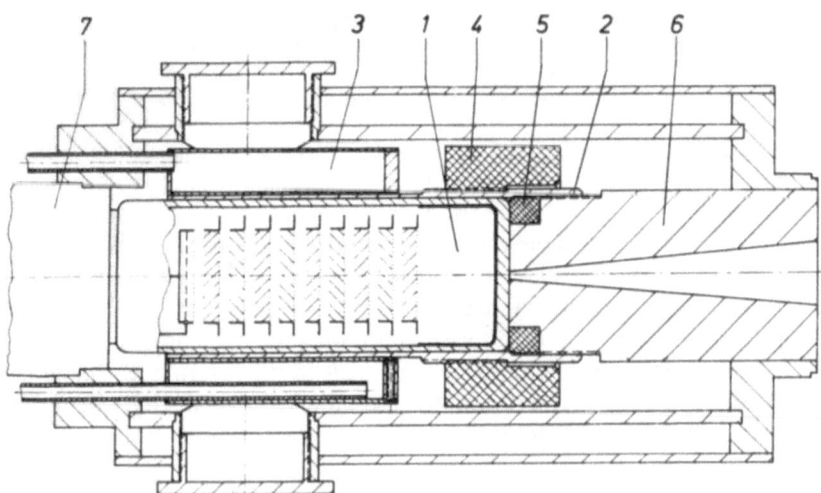

Fig. 2. Photomultiplier housing[18] with provisions for concentrating the light on the center of the photocathode, multiple reflections, magnetic defocussing, and cooling. 1) EMI 9558 QA photomultiplier; 2) brass tube; 3) stainless-steel cylinder for coolant; 4) coil; 5) permanent ring magnet; 6) lucite cylinder with cone; 7) socket.

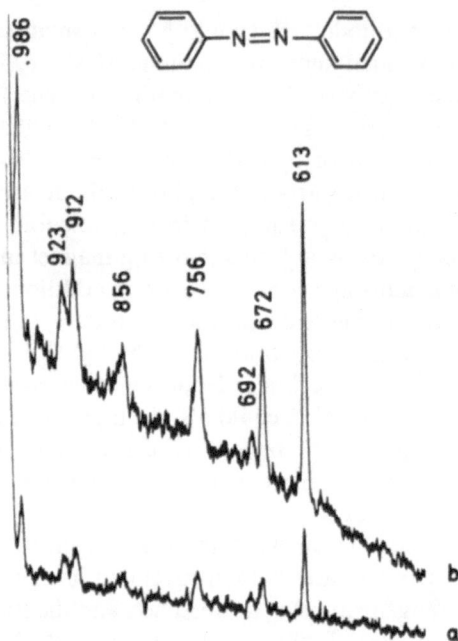

Fig. 3. Demonstration of S/N improvement by magnetic defocussing. Sample: polycrystalline azobenzene. a) Cooled RCA 7265 photomultiplier; b) uncooled EMI 9558 QA with magnetic defocussing. Slit, $s = 3\,\text{cm}^{-1}$; scanning speed, $w = 50\,\text{cm}^{-1}/\text{min}$; time constant, $\tau = 2\,\text{sec}$.

recorded without magnetic defocussing with a cooled photomultiplier, and spectrum b with magnetic defocussing, but without cooling. The increase in signal-to-noise ratio by a factor of three is evident.

Further improvement is possible through photon counting as the considerations above have shown. Figure 4 shows the schematic arrangement of such a system as described by Topp and coworkers.[18] The negative pulses of 50-nsec duration coming from the voltage follower are preamplified to surpass the lowest triggering threshold of the discriminator. The threshold is so adjusted that the dynode noise pulses are too small to trigger the monostable multivibrator of the discriminator.

The dead time of the discriminator is only 100 nsec. The observed count rate of a statistical process is[25]

$$Z_{\text{obs}} = \frac{Z}{1 + \tau Z}$$

where Z is the true count rate and τ the dead time of the counter. This formula gives an error in the intensity ratio of 2 Raman lines with 100,000 and 50,000 Hz, respectively, of 0.5 % for a dead time of 100 nsec and of more than 20 % for $\tau = 3$ μsec, a value measured for an often-used commercial ratemeter.

The discriminator is followed by a number of flip-flops as a binary divider. The divider gives 13 ranges with $2^7 = 128$ to $2^{19} = 524,288$ Hz for full-scale deflection of the recorder. A monostable multivibrator produces a well defined rectangular pulse from every negative slope coming from the divider. For ranges of $2^3 = 8$ to $2^6 = 64$ Hz the pulse length is increased accordingly. The integrator averages the charges of the pulses and gives at the output a dc signal which is proportional to the number of pulses at the input. Instead of a simple RC combination an RCL combination is used for the integration, because this better suppresses high frequencies which do not contribute information to the Raman signal. Up to the cut off frequency of 0.64 Hz the frequency characteristics of the integrator correspond to that of a low pass filter of $RC = 0.25$ sec.

With the photon counting system the effectiveness of the magnetic defocussing was tested. The dark count rate of the uncooled EMI 9558 QA photomultiplier (supply voltage, 2000 V) dropped from 6700 Hz to 300 Hz with the coil alone, and to 150 Hz with the permanent ring magnet. Cooling to $-30°C$ brought a further decrease to less than 5 Hz.

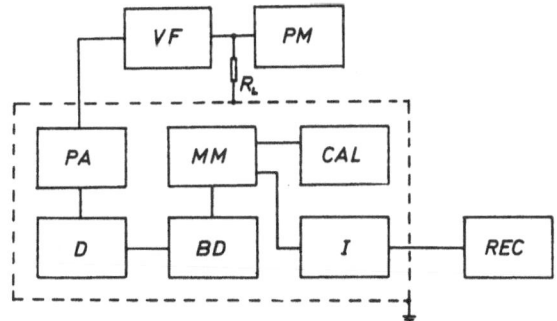

Fig. 4. Block diagram of a photon counting system.[18] PM) Photomultiplier; R_L) load resistance; VF) voltage follower; PA) preamplifier; D) discriminator; BD) binary divider; MM) monostable multivibrator; CAL) calibration; I) integrator; REC) recorder.

Figure 5 demonstrates the final improvement through pulse counting and magnetic defocussing. The spectra of crystalline azobenzene were recorded with a Raman spectrometer built around a Jarrell–Ash double monochromator, which will be described in the next section. Spectrum *a* was recorded with the dc system with cooling, and spectrum *b* with magnetic defocussing, photon counting, and cooling. The signal-to-noise ratio is sixfold in *b* over *a*.

Figure 6 shows a survey spectrum of polycrystalline azobenzene recorded with a scanning speed of 200 cm^{-1}/min and a response time of 0.25 sec. For the lattice vibrations resolved to 10 cm^{-1} the slit was closed to 2 cm^{-1}. At left the lines corresponding to the CH vibrations are shown in spite of the much smaller spectral sensitivity at longer wavelengths. The upper trace is recorded on a fourfold scale to show the details more clearly. This region could previously not be recorded with good resolution.

For intensity measurements, and when the power of the light source is strongly fluctuating, it is necessary to record only the ratio of the

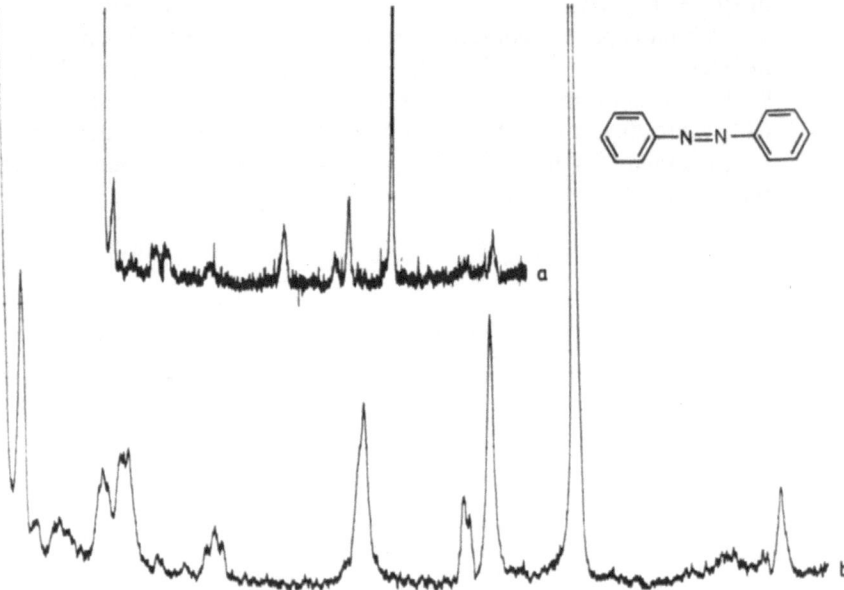

Fig. 5. Demonstration of S/N improvement by magnetic defocussing and photon counting. Sample: polycrystalline azobenzene, 450 to 1000 cm^{-1}. a) Cooled RCA 7265 photomultiplier with dc detection; b) cooled EMI 9558 QA with magnetic defocussing and photon counting. $s = 2$ cm^{-1}, $v = 10$ cm^{-1}/min, $\tau = 4$ sec.

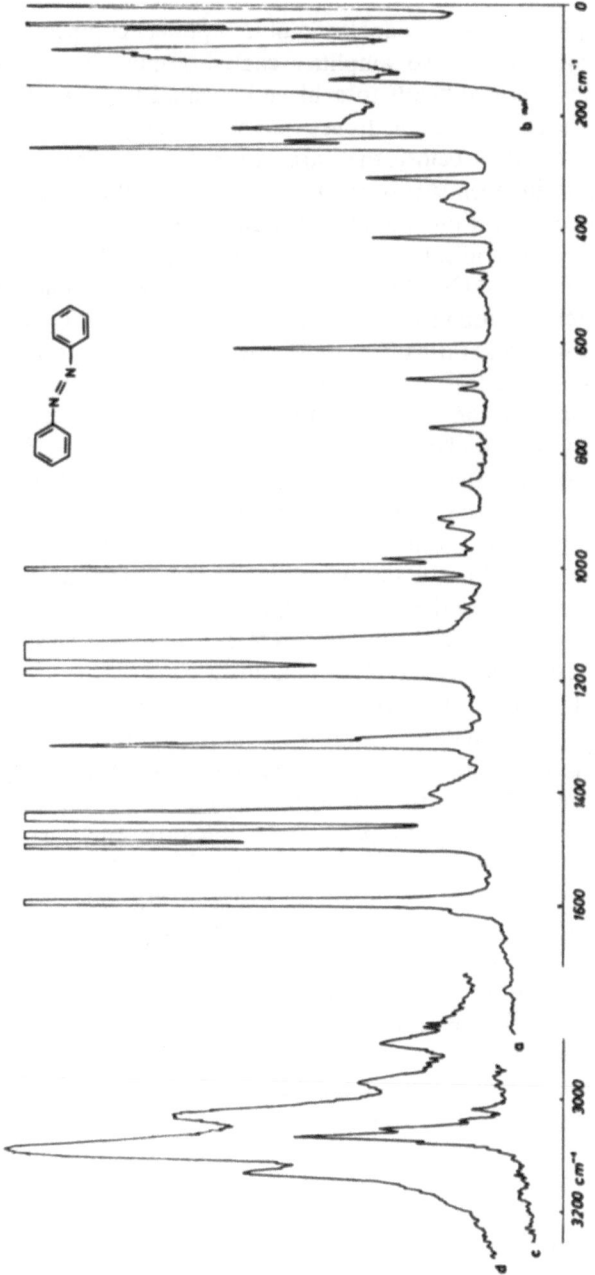

Fig. 6. Survey spectrum of polycrystalline azobenzene. a) $s = 3$ cm^{-1}, $v = 200$ cm^{-1}/min, $\tau = 0.25$ sec, range, 8000 Hz; b) $s = 2$ cm^{-1}, $v = 100$ cm^{-1}/min, $\tau = 0.25$ sec; c) $s = 3$ cm^{-1}, $v = 20$ cm^{-1}/min, $\tau = 4$ sec, range, 1000 Hz; d) $s = 3$ cm^{-1}, $v = 5$ cm^{-1}/min, $\tau = 8$ sec, range, 500 Hz.

Raman signal and a reference signal. To obtain perfect compensation of rapid fluctuations the two amplifier channels must have equal response times. This is difficult to achieve for small time constants. In a photon counter it is possible to use the reference signal to change the height of the pulses before they are integrated.[104] This makes the compensation independent of the time constant of the integrator.

Apart from dc detection and photon counters many workers have used lock-in detectors and some the signal-in-noise method of Pao and Griffiths.[26,1] Alfano and Ockman[27] have compared these systems with a photon counter and found for a certain weak signal for the lock-in detector a signal-to-noise ratio of 8.2, for the signal-in-noise detector 27, and for the photon counter 32. As already became clear from the above considerations the photon counter is the superior detection system. Kozlov[28] has, moreover, claimed that the signal-in-noise method can, in principle, never be better than a very good dc detection system.

In connection with the quasicontinuous ruby laser, photon counting systems cannot be used because one laser spike may excite more than one Raman photon within the resolving time of the system, and nonlinearity would result. On the other hand the emission characteristics of the laser with pulses of 1 msec duration at 50 Hz allows a suppression of the dark current by a factor of 20 with a linear gate.

We have developed an electronic detection system[29] for quasicontinuous ruby laser excitation, shown schematically in Fig. 7. The signal from the photomultiplier is amplified and fed into the linear gate. The gate is triggered by a pulse generator which is synchronized with the current of the laser flashlamp. A linear integrator converts the signal pulses to a dc signal which is recorded by a Speedomax-G recorder. The quasicontinuous ruby laser is not a very stable source. To avoid random fluctuations of the recorded signal it is necessary to record only the ratio of the Raman signal to a reference signal. A flexible light pipe directs scattered light from the interference filter onto a reference multiplier (Hamamatsu R 136). The preamplified signal is converted to dc by an identical linear integrator, the output of which serves as reference voltage for the recorder.

Figure 8 shows the improvement of the signal-to-noise ratio by this detection system. The Raman spectrum of carbon tetrachloride serves as an example. Spectrum Ia was recorded without and spectrum Ib with activation of the gate. The v_3 doublet at 762–790 cm^{-1} was also recorded with increased gain—IIa without, IIb with gate. While the signal height in spectra a and b is about equal, the electronic back-

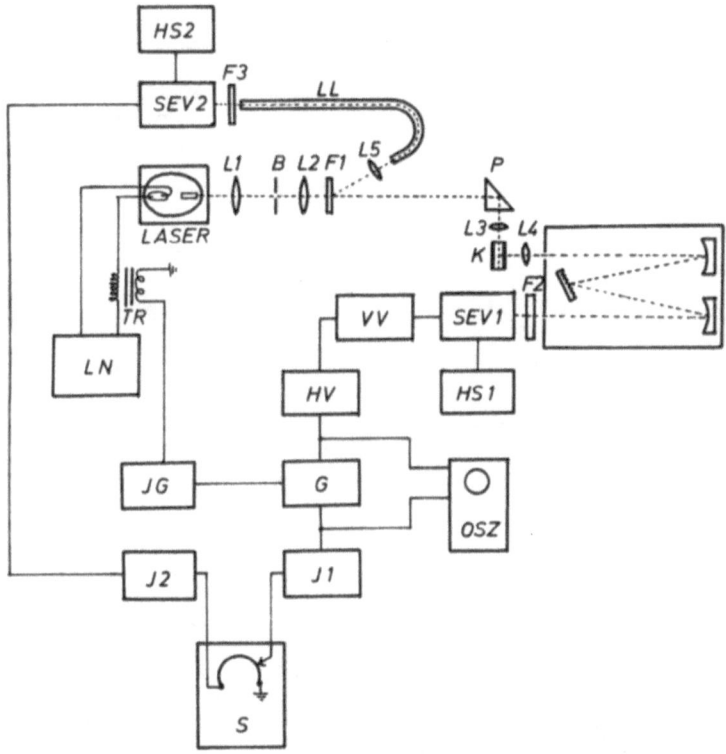

Fig. 7. Block diagram of electronic detection system for quasicontinuous ruby laser excitation. L) Lenses; B) diaphragm; F) filters; P) prism; K) sample cell; LL) flexible light pipe; SEV) photomultipliers; HS) stabilized high-voltage power supplies; VV) preamplifier; HV) main amplifier; G) gate; TR) transformer; LN) laser power supply; JG) pulse generator; OSZ) oscilloscope; J) integrators; S) recorder.

ground is lower by about a factor of 20 in spectrum b (both spectra were recorded without zero suppression) and the signal-to-noise ratio consequently improved by a factor of $\sqrt{20} \approx 4.5$.

A further possibility for improving the signal-to-noise ratio in Raman spectra is multiple scanning and storing of the signals in a time-averaging computer. With n-fold scanning the signal-to-noise ratio is improved by a factor of $\sqrt{n}$. Ziegler and Hoffmann[30] have shown that even 15-fold scanning decisively improves the information obtained from a noisy signal.

The next step is direct data storage and processing with a digital computer. Corrections of the spectra for instrumental influences, such

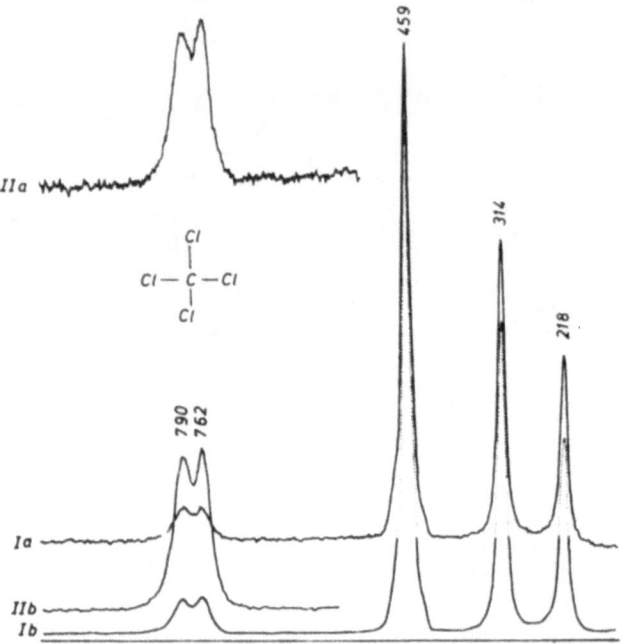

Fig. 8. Demonstration of S/N improvement by the linear gate. Raman spectrum of carbon tetrachloride. Ia: dc detection; Ib: with gate, $s = 5\,\mathrm{cm}^{-1}$, $\tau = 1\,\mathrm{sec}$, $v = 90\,\mathrm{cm}^{-1}/\mathrm{min}$; II$a$: dc detection, II$b$: with gate, gain fivefold in II.

as slit function, spectral sensitivity, and instrument polarization, can be made automatically, and the results of measurements, such as line position, integrated intensity or scattering coefficient, depolarization ratio, line width, etc., are obtained as printout of the computer. The technique has been applied by Scherer[105] and Murphy[106] for Raman spectroscopy.

Complete Raman Spectrometers

The first commercially available Raman spectrometer with laser excitation was the Perkin–Elmer LR-1.[31] It has recently been equipped with a postmonochromator and a linear wave-number drive[32] and is now called LR-3.

The Cary 81 which was originally constructed for mercury lamp excitation and employs a Littrow double monochromator was converted to He–Ne laser excitation.[1,31] For better accuracy of depolarization-ratio measurements a multiple reflection cell can be used.[33]

The low intensity of Raman scattering vs the stray light of many samples soon called for the general use of double monochromators. Landon and Porto[20] coupled two single monochromators to form a tandem instrument. From this start the Spex 1400 (and 1401) double monochromator was developed which has been used for most of the pioneer work on the Raman spectra of crystals. It consists of two Czerny–Turner monochromators with 80-cm focal length and additive dispersion. The two gratings are coupled and driven by a common lead screw, in the model 1401 linearly in wave numbers.

The Spex "Ramalog" Raman spectrometer employs in its sample compartment the principles of focussing and scattered light collection treated above. All provisions for accurate depolarization measurements are made. A quartz wedge in front of the slit serves as polarization scrambler to avoid measurement errors due to the instrument polarization.

An ITT FW-130 photomultiplier in a thermoelectrically cooled housing is used as detector. The dc amplification system can be supplemented by a photon counter for applications with extremely low light levels.

The Coderg PH 1 Raman spectrometer is constructed around an Ebert double monochromator with 60-cm focal length and additive dispersion. Figure 9 schematically shows the arrangement of the

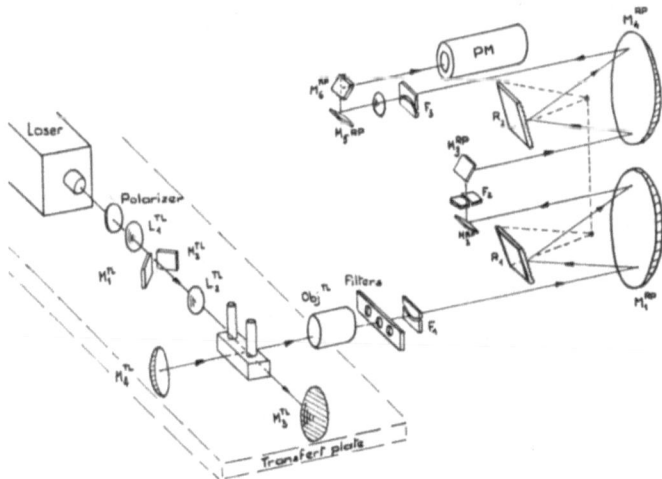

Fig. 9. Coderg Raman spectrometer PH 1. L) Lenses; M) mirrors; F) slits; R) gratings; PM) photomultiplier.

instrument. The laser beam is focussed into the sample cell and multiply reflected between a concave mirror and two plane mirrors arranged prismatically. The scattered light is concentrated on the slit by an objective lens; a concave mirror also collects the backscattered light. The curved slits are coupled and their widths can be changed in steps of 1:2. By the choice of slit width w and scanning speed v, the time constant τ is adjusted so that the optimal recording conditions according to Schubert's[34] formula $v\tau = w/4$ are always maintained.

The sample holder plate shown in Fig. 9 is used for liquids and gases. Special accessories are available for microsamples, crystal powders, oriented crystals, and low-temperature work.

The performance of the instrument is best illustrated by a recording of the vibration–rotation spectrum of oxygen in air shown in Fig. 10. The photon counter used for recording this spectrum was optional and is now incorporated into the new model PH O. Magnetic defocussing is always applied.

Coderg also constructed a smaller tabletop Raman spectrometer "Automatique" specially designed for routine analysis in chemical

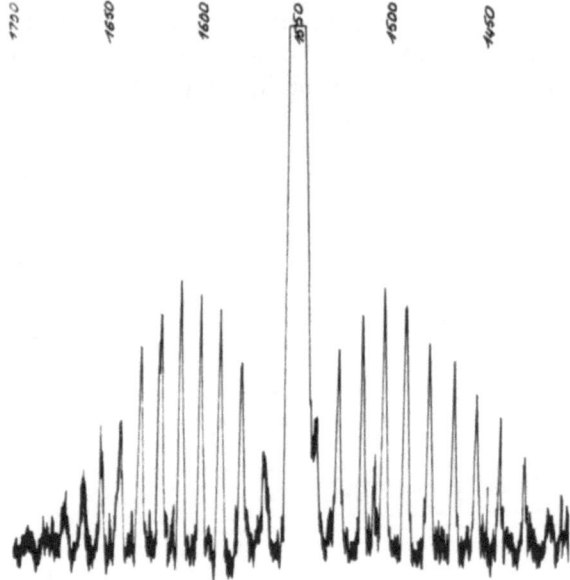

Fig. 10. Vibration–rotation band of O_2 in air (0.21-atm partial pressure) excited with a power of 500 mW at 488 nm. $s = 2\,\mathrm{cm}^{-1}$, $v = 3\,\mathrm{cm}^{-1}$/min, range, 200 Hz.

laboratories. A Spectra-Physics model 124 He–Ne laser serves as light source; the Ebert double monochromator has a focal length of 30 cm. Figure 11 shows the optical design of the instrument. This instrument has been redesigned by Spectra-Physics into their model 700 Raman spectrometer.

The latest complete Raman spectrometer has been developed by Jarrell–Ash. It is built around the double Czerny–Turner monochromator 25–100 with a 1-m focal length. The two gratings of 10×10 cm are arranged on a common pivot axis to achieve perfect tracking. In this design the second monochromator does not contribute to the dispersion.

Figure 12 schematically shows the arrangement of this Raman spectrometer. In the lower part of the instrument, room is provided for any commercial gas laser. The laser beam is deflected three times by Brewster angle prisms before it is focussed into the sample. The scattered light is collected by a lens with high light-gathering power. The alternate light paths allow the use of the transillumination technique[35] or irradiation from below, e.g., for low-temperature experiments in a Dewar. The sample compartment is big in order to accommodate many experimental arrangements without limitation of space. The three slits are coupled and a servosystem keeps them at a constant spectral slit width over the whole range of scanning.

The signals are detected by a thermoelectrically cooled ITT photomultiplier FW 130 and recorded by a photon counting system. A

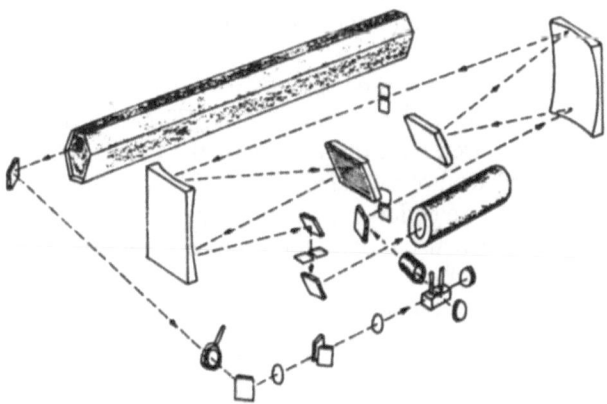

Fig. 11. Design of Coderg Raman spectrometer "Automatique."

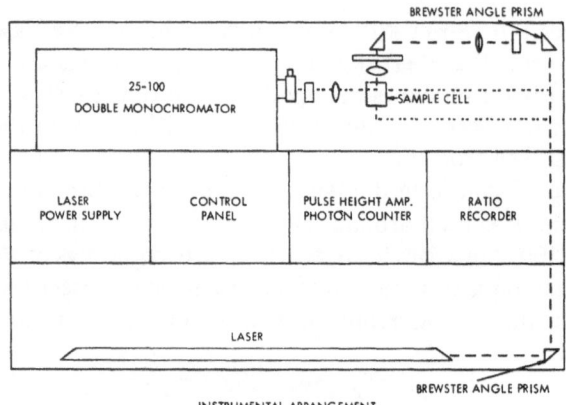

Fig. 12. Jarrell–Ash Raman spectrometer.

reference voltage proportional to the laser power is applied to the slide-wire of the recorder in order to compensate intensity fluctuations.

The performance of the instrument is demonstrated by part of the Raman spectrum of benzene in the region from 2400 to 3200 cm^{-1} in Fig. 13. At a full-scale sensitivity of 100 Hz the noise corresponds to about 3 Hz. The very weak combination lines at 2758 and 2783 cm^{-1} give a full-scale throw and previously unknown lines are detected at 2578, 2605, and 3147 cm^{-1}.

In Munich Brandmüller and coworkers[36] have built up a combination of two double monochromators and three lasers as a multipurpose Raman spectrometer. Figure 14 schematically shows the arrangement. The Jarrell–Ash double monochromator can be used either with the argon ion laser or with the He–Ne laser as source, and is equipped with a dc detection system with compensation for source fluctuation[24,37] and with the photon counter[18] described above. The RSV double monochromator can be used with the He–Ne or the quasicontinuous ruby laser, and the spectra are recorded either with a commercial photon counter or with the detection system developed for the ruby laser (see above). The latter can also be used for dc detection when the gate is not activated. Broadband dielectric mirrors are used to reflect the laser beams in the desired directions.

Michel[101] describes a Raman spectrometer with a Czerny–Turner double monochromator of his own construction with a focal length of 2 m, arranged side by side to achieve additive dispersion. It is equipped with two gratings (154 × 206 mm^2 with 600 grooves/mm) which are

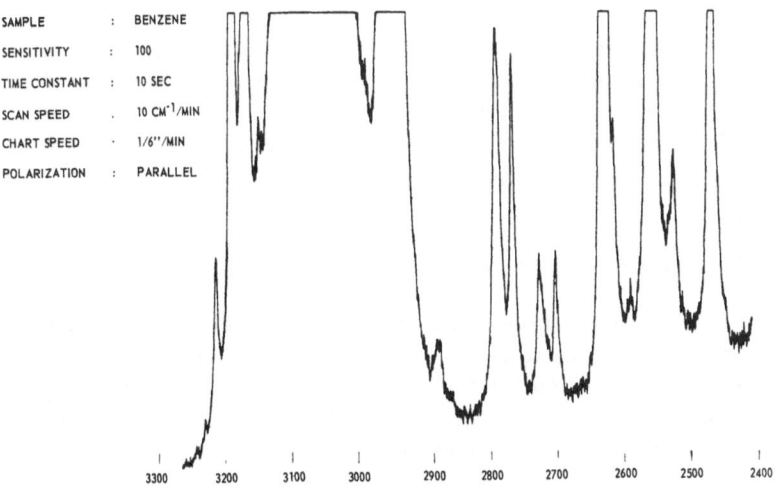

SAMPLE	:	BENZENE
SENSITIVITY	:	100
TIME CONSTANT	:	10 SEC
SCAN SPEED	.	10 CM⁻¹/MIN
CHART SPEED	·	1/6"/MIN
POLARIZATION	:	PARALLEL

Fig. 13. Raman spectrum of benzene from 2400 to 3250 cm^{-1}. Slit, 100 μm.

used in second order. An OIP He–Ne laser with 150 mW multimode or a mercury arc is employed for excitation. The spectra which are recorded with dc detection have an excellent signal-to-noise ratio.

Delhaye[12] has modified a Coderg PH 1 spectrometer for rapid scanning with 1000 cm^{-1}/sec. The spectra are displayed on an oscilloscope screen and filmed or recorded on magnetic tape. In this way chemical reactions can be studied through the changes in the Raman spectrum. For fast reactions Delhaye[12] employs his electro-optical

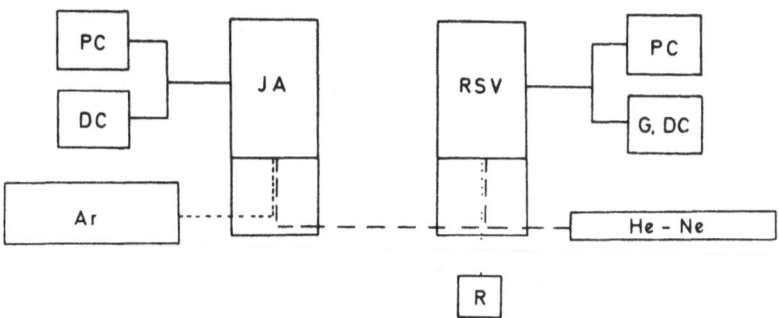

Fig. 14. Block diagram of Raman spectrometers in Munich.[36] JA) Jarrell–Ash 25–100 double monochromator; Ar) Spectra-Physics model 140 argon ion laser (-----); RSV) double monochromator; He–Ne) Spectra-Physics model 125 He–Ne laser (————); R) Siemens quasicontinuous ruby laser (·····); PC) photon counter; DC) detection system; G) gate electronics.

spectrographs with image intensifier phototubes which allow the recording of a complete Raman spectrum over a range of a few hundred cm^{-1} with a single flash of a pulsed laser. With the quasicontinuous ruby laser[11] 50 spectra/sec can be recorded. Coderg has announced the commercial production of such an electro-optical spectrograph.

METHODS AND RESULTS FOR AMORPHOUS MEDIA

As already mentioned, it is not possible to include all results of Raman spectroscopy with lasers in this review. In this second part the methods of measurement for wave numbers, depolarization ratios, intensities, and line widths will be described together with examples of results obtained with these methods.

Measurement of Wave Numbers

The measurement of the wave-number shifts v of the Raman lines from the exciting line v_0 is based on a measurement of wavelengths. Raman spectrometers are usually calibrated with emission lines from rare-gas discharge lamps which are well tabulated.[38] Since these lines are also emitted by gas-laser discharges they can often be directly used for calibration.

Because wavelengths are measured in air and wave numbers in vacuum, the conversion formula contains a correction term δ:

$$v = v_0 + \delta - 1/\lambda_{air}$$

The values of δ have been tabulated by Opler[39] and recently by Strey[40] for an extended region of the spectrum including laser-excited Raman spectra. Direct conversion tables from wavelengths to Raman wave-number shifts are also obtainable[40] for exciting lines of He–Ne, argon, and ruby lasers.

Most modern Raman spectrometers record the spectra on a linear wave-number scale. However, in fact this is often an inverse wavelength scale, and for accurate measurements the δ correction should be applied.

The Raman spectrometers with He–Ne laser excitation have widely extended the number of substances for which the Raman spectra can be recorded, because absorption and fluorescence are not as much a problem at 632.8 nm as at 435.8 nm, the most frequently used mercury lamp exciting line.

For instance, Burchardi[41,24] has investigated the Raman spectra of 18 azobenzene derivatives and based the assignment of the lines on a molecular model of azobenzene calculated with the well known FG matrix method.[42] In such calculations the knowledge of the wave numbers in the Raman spectra of isotopically substituted molecules is of great help. Therefore, Burchardi has also recorded the spectrum of azobenzene—$^{15}N_1$, which is shown in Fig. 15 together with the spectrum of normal azobenzene.[24] The greatest shift is observed for the line at 1441 to 1429 cm^{-1}, which corresponds to the N=N bridge vibration, while the line at 1492 cm^{-1} loses a great fraction of its intensity, a fact which will be discussed in a later section.

It was intended to improve the calculation of the molecular force field of azobenzene by the data from the spectra of azobenzene-$^{15}N_2$, azobenzene-d_{10}, and p-dideutero-azobenzene recorded with the Raman spectrometer described above.[36] However, it turned out that all observed lines were polarized, which indicates that azobenzene is not a planar molecule,[43] and that further studies of the problem are required.

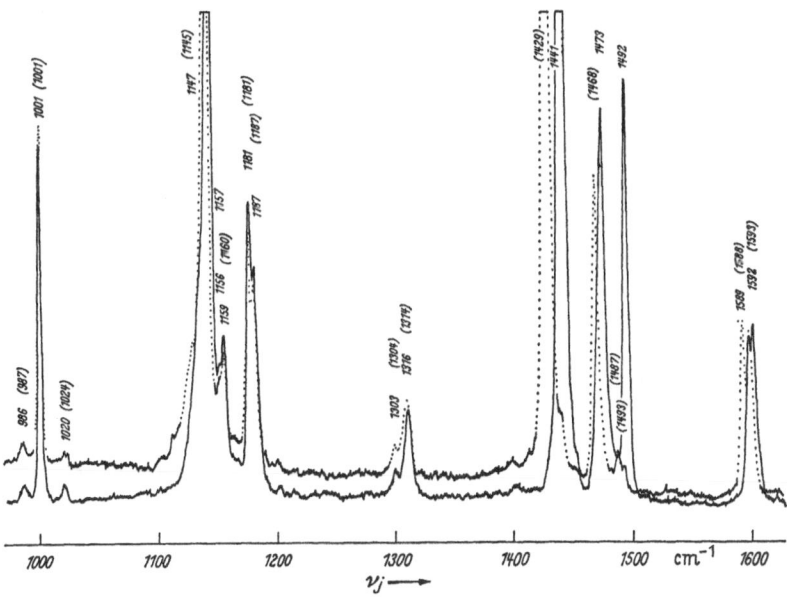

Fig. 15. Raman spectrum of polycrystalline azobenzene (———) and azobenzene-$^{15}N_1$ (·····, wave numbers in parentheses).[24] Pellet thickness, 0.2 mm; $s = 3$ cm^{-1}; $\tau = 2$ sec; $v = 12$ cm^{-1}/min.

Measurement of Depolarization Ratios

The second important quantity in a Raman spectrum is the depolarization ratio ρ_s of a line, defined as the ratio of the intensities $I_{\parallel}$ for exciting light linearly polarized parallel to the direction of observation and $I_{\perp}$ for exciting light polarized perpendicular to it.[44] It is obvious that the linearly polarized light of gas lasers allows a very accurate and convenient measurement of depolarization ratios.

A simple example is shown in Fig. 16.[24] Of the three lines of carbon tetrachloride one is completely polarized ($\rho_s \approx 0$) and the other two are depolarized ($\rho_s = \frac{3}{4}$). These values are directly obtained with method (a) when the direction of polarization of the incident exciting light is changed and an analyzer transmits only light polarized perpendicular to the spectrometer slit. When the direction of polarization of the exciting light is kept in the perpendicular position and the analyzer in front of the slit is turned instead (b), the observed depolarization ratios must be corrected for the instrument polarization.[45] This cumbersome procedure may be avoided by a polarization scrambler, e.g., a quartz wedge,[45] between analyzer and slit. Lau[46a] and Claassen

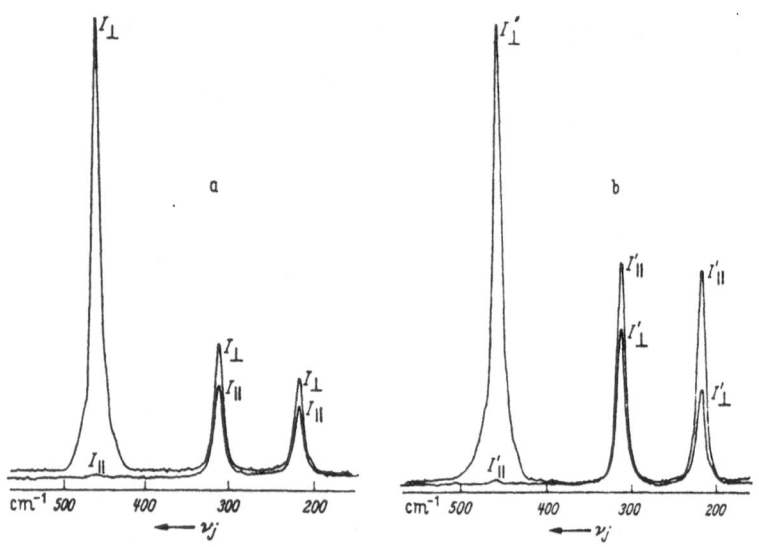

Fig. 16. Depolarization-ratio measurement in the Raman spectrum of carbon tetrachloride,[24] a) by turning the polarization rotator and b) by turning the analyzer. $s = 5 \text{ cm}^{-1}$, $v = 50 \text{ cm}^{-1}/\text{min}$, $\tau = 1$ sec.

et al.[46b] propose to rotate the polarization of the scattered light by 90°
because the efficiency of many gratings is higher for light polarized
parallel to the rulings for wavelengths shorter than the blaze wavelength. They discuss the eight resulting methods to measure depolarization ratios with and without analyzer, scrambler, rotator, and combinations thereof. Allemand[47] has tested the accuracy of these eight methods
and found that method (a) gives the best results for depolarized lines
and therefore should be used to distinguish between depolarized and
weakly polarized lines (the use of an analyzer, however, significantly
reduces the available intensity, and for weak lines one of the methods
which avoids the analyzer may be more advantageous).

As Bridge and Buckingham[17] have shown, method (b) is, on the
other hand, preferable for the precise measurement of small depolarization ratios, because the effects of the collection angle of the scattered
light on the observed ρ_s are much smaller than for method (a). This is
due to the dependence of the Raman intensity on the direction of observation which has been studied by Porto and coworkers[48,1] and Skinner
and Nilsen.[49] They obtained good agreement with theoretical expectation, so that it is only necessary to measure the depolarization ratio in
order to know the angular dependence of the intensity of a particular
line.

Hess and coworkers[50] have calculated the systematic error of
depolarization measurements with a finite collection angle 2φ from the
angular dependence of the Raman intensity. They obtain for method
(a),

$$\rho_s(\varphi) = \rho_s(0) + \frac{\varphi^2}{3} \cdot (1 - \rho_s^2) + \cdots$$

where $\rho_s(\varphi)$ is the observed and $\rho_s(0)$ the true depolarization ratio. For
$\rho_s = \frac{3}{4}$ the deviation is negligible; for very small ρ_s, however, $\varphi^2/3$ gives
a considerable error. For $\rho_s = 0.007$ and $\varphi = 3°$ the error is 13%, and
for $\varphi = 5°$ as much as 36%.

For this reason method (b) is preferable for the measurement of
small ρ_s. In this case Hess[50] obtained

$$\rho_s(\varphi) = \rho_s(0) + \rho_s\frac{\varphi^2}{3} \cdot (1 - \rho_s) + \cdots$$

Now the correction term decreases with ρ_s and becomes negligible.
This agrees essentially with Bridge and Buckingham's[17] result. They
have, moreover, shown that the influence of the convergence angle of the
incident laser beam on small ρ_s is only negligible ($<10^{-4}$) for angles

$< 1°$, which, for example, requires a focussing lens of $f \gtrsim 100$ mm for a laser beam of $d \approx 2$ mm.

Hess and coworkers[50] have measured the depolarization ratios of the Raman lines of a number of liquids. The results are presented (together with their intensity data) in Table 1.

Table 1. Relative Corrected Scattering Coefficients S and Depolarization Ratios ρ_s[50]

ν (cm^{-1})	S	ρ_s	ρ_s[52]
		Carbon tetrachloride, CCl$_4$	
218	0.16 ± 0.01	0.75 ± 0.02	
314	0.31 ± 0.01	0.75 ± 0.02	
459	1.00	0.0075 ± 0.0015	0.0060 ± 0.0005
762⎫ 790⎭	0.66 ± 0.02	0.72 ± 0.02	
1535	0.20 ± 0.01	0.15 ± 0.01	

ν (cm^{-1})	S	ρ_s	ν (cm^{-1})	S	ρ_s
	Chloroform, CHCl$_3$			Dichloromethane, CH$_2$Cl$_2$	
262	0.155	0.75	283	0.113	0.39
366	0.26	0.09	704	0.71	0.06
668	0.63	0.01	738	0.205	0.58
761	0.335	0.72	1148	0.10	0.74
1215	0.18	0.71	1420	0.32	0.71
3018	2.68	0.1	2985	4.70	0.06
			3048	1.29	0.75
	Benzene, C$_6$H$_6$				
				Cyclohexane, C$_6$H$_{12}$	
606	0.128	0.75			
850	0.062	0.62	385	0.021	0.15
992	3.52	0.02	429	0.035	0.75
1178	0.51	0.72	802	1.17	0.07
1585⎫ 1604⎭	0.90	0.75	1030	0.85	0.75
			1159	0.14	0.22
2910	1.10	0.28	1267	0.97	0.74
3045⎫ 3064⎭	20.2	0.22	1349	0.18	0.75
			1445	1.30	0.72
3286	0.9	0.1	2634	0.23	0.1
			2666	1.02	0.3
	Carbon disulfide, CS$_2$		2699	0.81	0.3
648⎫ 656⎭	1.87	0.135	2852⎫ 2893 2924⎬ 2939⎭	55.5	0.19
796⎫ 805⎭	0.38	0.09			

Table 1 (continued)

$\nu(\text{cm}^{-1})$	S	ρ_s	$\nu(\text{cm}^{-1})$	S	ρ_s
Methanol, CH_3OH			Ethanol, C_2H_5OH		
1034	0.28	0.15	882	0.27	0.13
1112 ⎱ 1164 ⎰	0.07	0.3	1047 ⎱ 1095 ⎱ 1115 ⎰	0.38	0.31
1450 ⎱ 1470 ⎰	0.152	0.5	1274	0.082	0.52
2835 ⎱ 2913 ⎱ 2944 ⎱ 2989 ⎰	9.5	0.11	1454 ⎱ 1480 ⎰	0.60	0.62
			2876 ⎱ 2928 ⎱ 2974 ⎰	16.3	0.13

For totally symmetric vibrations of molecules of cubic point groups the depolarization ratio should be zero. For the a_1 line of carbon tetrachloride small values of ρ_s have always been measured, but definitely not zero. Chantry[51] has shown that this discrepancy is not due to an isotope effect. The high power of the argon ion laser has made it possible to measure ρ_s of this line with higher accuracy and also at concentrations of 10 % in various solvents.[52] The drastic variations of the depolarization ratio with the solvents suggest that the deviation from zero is due to intermolecular interactions. This was already concluded from earlier measurements with He–Ne laser excitation.[53]

It has also become possible to accurately measure depolarization ratios of very weak lines, e.g., overtone and combination lines. Figure 17 shows, as example, the region from 1650 to 3250 cm^{-1} of the Raman spectrum of benzene.[102] The instrument polarization makes the depolarization ratio of depolarized bands appear to be smaller than $\frac{3}{4}$ because method (b) was used. The spectrum was excited with an argon ion laser (Spectra-Physics model 140, 600 mW at 488 nm) and recorded with a Cary 81 Raman spectrometer which was modified in a similar manner as described by Walrafen.[54] The band at 3062 cm^{-1} was recorded with sensitivity $S = 3.2$, the bands at 2950 and around 3189 cm^{-1} with $S = 20$, and the region from 1650 to 2900 cm^{-1} with $S = 100$, repeated from 1650 to 2400 cm^{-1} with $S = 500$. It is obvious that the lines at 2298 and 2458 cm^{-1} are polarized, while Sushchinskii and Muldakhmetov[55] reported them to be depolarized and, consequently, assigned them to combinations of species E_{1g}. The polarized line 2618 cm^{-1} was correctly assigned to the overtone of the forbidden b_{2u}

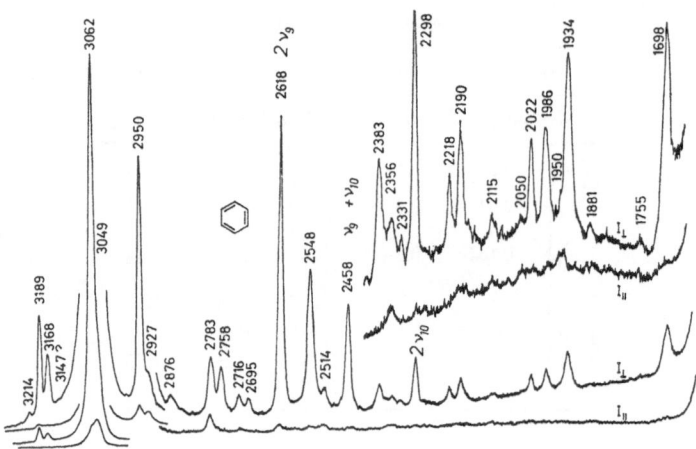

Fig. 17. Depolarization-ratio measurement in the Raman spectrum of benzene.[102] $s = 10\,cm^{-1}$, $v = 30\,cm^{-1}$/min, $\tau = 5$ sec (10 sec for upper two traces). For sensitivities see text.

vibration v_9 which was placed at $1309\,cm^{-1}$ by Mair and Hornig.[56] Therefore, it seems to be appropriate to assign the two lines at 2298 and $2458\,cm^{-1}$ to the overtone of v_{10} (the other b_{2u} vibration at 1150 cm^{-1}) and to the combination $v_9 + v_{10}$. The consistent assignment of these three lines to overtones and a combination of the inactive b_{2u} vibrations v_9 and v_{10}, gives further decisive support to the long disputed Mair and Hornig assignment of v_9. The investigation of the Raman spectrum of hexadeuterobenzene confirmed this result.[107] The lines at 1656, 2112, and $2573\,cm^{-1}$ are highly polarized and were assigned to $2v_{10}$, $v_9 + v_{10}$, and $2v_9$, respectively.

Intensity Measurements

Although the corrections and calibrations necessary for intensity measurements[57-59] have been considerably simplified by laser excitation, the results obtained to date with this method are relatively scarce.

A few direct scattering cross-section measurements[48,49,60,61] have been reported. Skinner and Nilsen[49] compare their result for the total differential cross section $d\sigma/d\Omega$ of the $992\,cm^{-1}$ line of benzene with those of the other workers. In order to allow for the different exciting lines used, they take the v^4 law into account. The values of $(d\sigma/d\Omega) \cdot (1/v^4) \cdot N$ are presented in Table 2 (N is the molecular density).

The agreement and accuracy of the data is obviously not yet very satisfactory.

Eckhardt and Wagner[62] relate the scattering cross sections to the conventional relative corrected scattering coefficients or "standard intensities"[57] S. However, because for liquids the scale factor for conversion to absolute scattering coefficients is still not accurately determined, the resulting cross sections are not more reliable than those given in Table 2.

Table 2. Total Differential Cross Section $d\sigma/d\Omega$ of the Raman Line 992 cm^{-1} of Benzene

Reference	60	48	61	49	
$\dfrac{d\sigma}{d\Omega} \cdot \dfrac{1}{v^4} \cdot N$	1.19 ± 0.6	0.94 ± 0.2	2.31 ± 1.2	1.78 ± 0.2	$\cdot 10^{-24} \dfrac{\text{cm}^3}{\text{sr}}$

For a determination of the gain coefficients of the stimulated Raman effect[3] the peak differential cross sections are of primary interest. They largely depend, however, on the accuracy of the measurements of the line widths which will be treated in a later section. Johnston and coworkers[63] have determined the gain coefficients for a number of materials from the peak differential cross sections.

Relative corrected scattering coefficients S for eight liquids have been measured by Hess and coworkers[50] with He–Ne laser excitation and are given in Table 1. In binary mixtures of CCl_4 with the investigated liquids they found no deviation from proportionality with the concentration.

Hacker[64] has treated the problem of interpretation and significance of Raman line intensities. He has measured relative scattering coefficients and depolarization ratios of 30 azobenzene derivatives and found an empirical relationship between the scattering coefficients and the potential energy distributions (PED), which gives a satisfactory description of the changes of intensity with isotopic substitution. The potential energy distribution P_{jk} is defined by [42]

$$P_{jk} = \frac{1}{(\Lambda_{\exp})_j} \left(\frac{\partial}{\partial f_k} \Lambda_j \right) \cdot f_k$$

where f_k is the force constant of bond or angle k and Λ_j is the eigenvalue of vibration v_j. Hacker[64] found that intensity changes on isotopic substitution (see Fig. 15 for an example) can be described empirically by the relation

$$S_j^{\frac{1}{2}} \sim \frac{\partial \alpha_{\rho,k}}{\partial Q_j} \sim f_1^{(k)} \chi_{\rho,k} \sum_{i \in (k)} L_{ij} P_{ij}^{\frac{1}{2}}$$

where j designates a normal vibration v_j with the normal coordinate Q_j, k is a group of the molecule with a defined polarizability where this vibration is localized (when such a localization is possible), α_ρ is a component of the polarizability in the molecular coordinate system, $f_1^{(k)} = v_e^{(k)}/(v_e^{(k)2} - v_0^2)$, with $v_e^{(k)}$ the wave number of an effective electron absorption band correlated to the polarizability α_k, a frequency factor originating from the theory of the resonance Raman effect,[65] χ_ρ is proportional to the oscillator strength of the effective electronic transition,[66] and L_{ij} are the components of the eigenvector. A rigorous derivation of this relationship was not possible to date, but it proves very useful as a working hypothesis. In Table 3 the calculated and observed scattering coefficients for three vibrations of the azo group in azobenzene, azobenzene-$^{15}N_1$, and azobenzene-d_{10} are compared. It should be noted that the wave numbers of azobenzene-d_{10} were calculated[41] without knowledge of the experimental values so that the force constants could not be adjusted for the best fit. In spite of this drawback the agreement between calculated and observed scattering coefficients for the lines v_{10} and v_{11} is excellent. Particularly, the change

Table 3. Comparison of Observed and Calculated Wave Numbers v_j and Scattering Coefficients S_j for Three Azo Group Vibrations in Azobenzene (a), Azobenzene-$^{15}N_1$ (b), and Azobenzene-d_{10} (c)

		v_j (cm^{-1}) obs.	v_j calc.	$S_j^{\frac{1}{2}}$ obs.	$S_j^{\frac{1}{2}}$ calc.
	a	1473	1490	6.3	3.0
v_7	b	1467	1489	5.4	2.9
	c	1514	1470	0.78	0.18
	a	1494	1436	5.0	4.8
v_{10}	b	1485	1429	2.0	1.8
	c	1342	1417	3.2	2.3
	a	1442	1410	8.8	8.1
v_{11}	b	1428	1395	10	10
	c	1467	1337	12	12

of intensity of the v_{10} line upon substitution of ^{15}N (see Fig. 15) is completely accounted for by the influence of the potential energy distribution.

Raman Spectra of Gases

The theory of Raman intensities is strictly valid only for gases because it is based on free orientation of the molecules. The influence of intermolecular interactions in liquids will be much better understood when more data measured in the gaseous phase become available for comparison. Therefore, it is of major importance that with the powerful argon ion lasers it is now comparatively easy to obtain good spectra from gaseous samples. With mercury excitation a sample volume of about 10 liters was needed.[67] Now a volume of a few ml or less is sufficient to obtain much better spectra than before.

A record in sample volume is held by Barrett and Adams.[15] They used the scattering from a volume of 10^{-8} cm^3 in the focus of a lens inside the cavity of an argon ion laser to record the vibration–rotation spectra of O_2 and N_2 shown in Figs. 18 and 19 with a Perkin–Elmer E-1 monochromator and a photon counting system. The high sensitivity

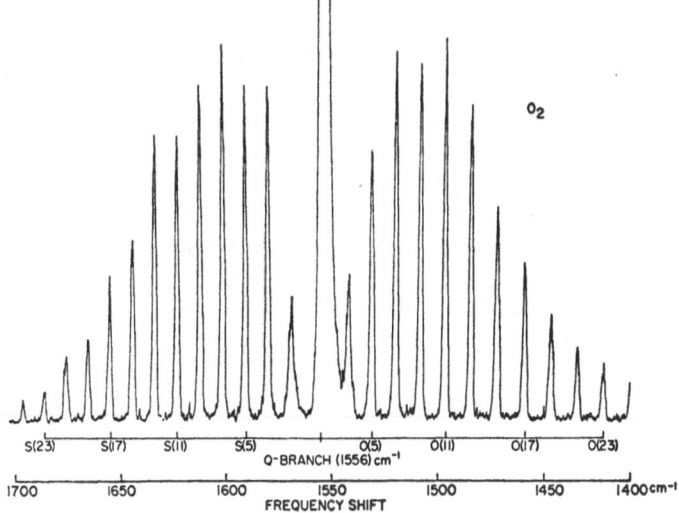

Fig. 18. $I_\parallel$ component of rotation–vibration band of O_2 at 1 atm, excited with a power of 3 W at 488 nm.[15] $v = 3.6$ cm^{-1}/min, $\tau = 3$ sec, $s = 20\,\mu m \approx 0.5$ cm^{-1}.[15]

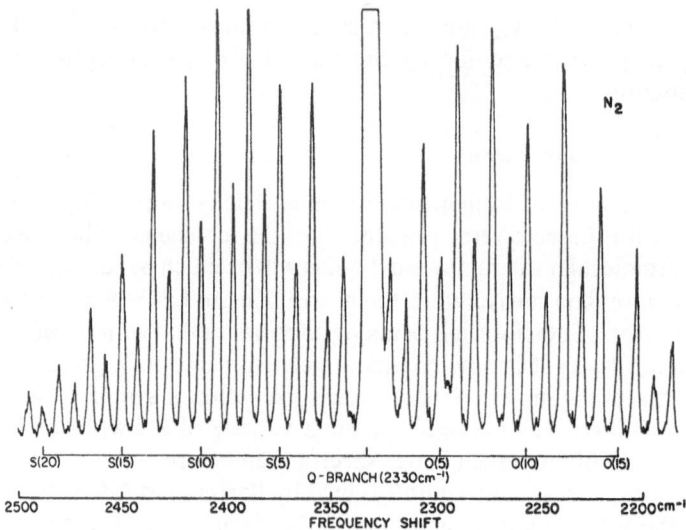

Fig. 19. $I_{\parallel}$ component of rotation–vibration band of N_2 at 1 atm, excited with a power of 3 W at 488 nm.[15] $v = 1.8\,\mathrm{cm}^{-1}/\mathrm{min}$, $\tau = 10\,\mathrm{sec}$, $s = 40\,\mu\mathrm{m} \approx 1\,\mathrm{cm}^{-1}$.[15]

of these experiments also made it possible to record the rotational Raman spectrum of vibrationally excited CO_2 in a gaseous discharge.[68] The rotational constants of the thermally populated level $v_1 v_2 v_3 = 01^1 0$ could be determined photographically.[108]

With high-power argon lasers, which deliver 1 W or more in one line, it is possible to obtain almost equivalent results outside the laser cavity. This can be recognized by comparison of Fig. 18 with Fig. 10, which shows the spectrum of O_2 in air excited by 500 mW.

The Raman spectra of vapors can now be recorded at low vapor pressures. For example, the spectrum of carbon tetrachloride vapor excited with $\sim 500\,\mathrm{mW}$ at 488 nm[69] is, at a pressure of 100 mm Hg, at least equivalent to that obtained earlier[70] at a pressure of 760 mm Hg by mercury excitation.

The intensity is high enough so that the slits can be narrowed to $1\,\mathrm{cm}^{-1}$ and the isotopic structure of the $459\,\mathrm{cm}^{-1}$ band of CCl_4 resolved in the spectrum of the vapor (see Fig. 21). It is even possible to lower the pressure to 5 mm Hg and still obtain a recognizable CCl_4 spectrum. From this pressure of 5 mm Hg one can estimate to what concentration a component in a mixture of liquids should be detectable in the Raman spectrum. As an example benzene in carbon tetrachloride was chosen.[69] Figure 20 shows the region around 1000 cm^{-1} in the spectrum of a solution of $5 \times 10^{-3}\%$ by volume of C_6H_6 in CCl_4. The line at 992 cm^{-1} of benzene is clearly seen. The detection limit is not so much determined by the sensitivity as by the background of weak combination lines of CCl_4. At about 1100 cm^{-1} a combination of the Fermi-resonance band v_3 with v_4 can be seen and in the slope of v_3 at $\sim 920 \text{ cm}^{-1}$ the overtone $2v_1$.

Holzer and coworkers[71] have observed a strong resonance Raman effect in Br_2, I_2, BrCl, ICl, and IBr vapor, when the exciting line falls into a region of continuous absorption of the gas. When the exciting line, however, coincides with a discrete energy level, fluorescence instead of Raman scattering is observed. The theory for these resonance effects has been developed by Behringer.[109] Kiefer[110] observed a superposition of a weak fluorescence and a resonance Raman effect in Br_2 for excitation with 694.3 nm.

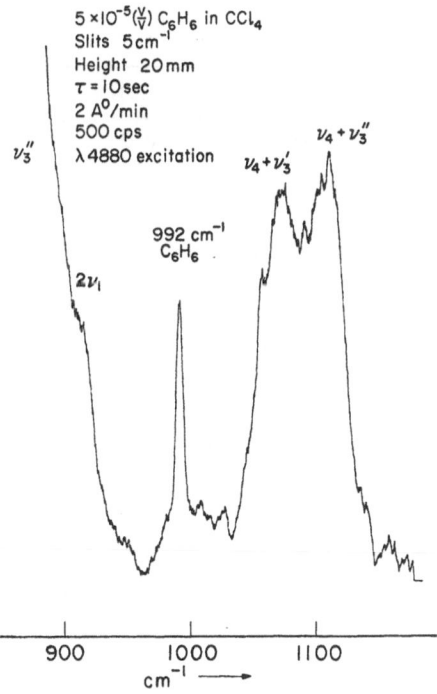

Fig. 20. Detection of $5 \times 10^{-3}\%$ by volume of benzene in carbon tetrachloride.

Measurement of Line Widths and Investigation of Band Profiles

Because the gain coefficients for the stimulated Raman effect[3] depend on the ratio of total scattering cross section to line width, much interest has been concentrated on an accurate determination of the width of Raman lines.

Clements and Stoicheff[72] have built an apparatus for this purpose consisting of a He–Ne laser of 4 m length which operates in a single longitudinal mode of $0.025 \, \text{cm}^{-1}$ half band width at 250 mW, a pressure-scanned Fabry–Perot interferometer, and a photon counting system. The Raman lines are isolated by interference filters. The measured line widths are given in Table 4 together with the data obtained by Scotto.[73] Earlier measurements with spectrometers gave values much too high for the narrower lines, e.g., for CS_2.

Table 4. Half Band Widths of Raman Lines of Liquids

		$\delta \, (\text{cm}^{-1})$	
Liquid	$\nu \, (\text{cm}^{-1})$	Ref. 72	Ref. 73
Liq. N_2	2331	0.067 ± 0.004	0.066 ± 0.006
Liq. O_2	1555	0.117 ± 0.008	0.135 ± 0.021
CS_2	656	0.50 ± 0.02	
Toluene	1003	1.94 ± 0.07	
Benzene	992	2.15 ± 0.15	

Since it became known that the line widths in mixtures are dependent on the solvent they are also of interest, because it is possible to study intermolecular interactions in this way.

Kroto and Pao[74] first observed this effect for the ν_1 band of carbon tetrachloride in various solvents. Figure 22 shows the changes of the line profile for solutions in cyclohexane, where the line width becomes smaller, and acetone and methanol, where it becomes greater. Murphy and coworkers[75] observed narrowing of the lines for solutions in carbon disulfide and benzene. The line profile in the vapor is shown in Fig. 21[69] for comparison. It may be concluded that the line broadening by neighboring molecules in the liquid grows in the order $CS_2 \lesssim C_6H_{12} \lesssim C_6H_6 < CCl_4 < C_3H_6O \approx CH_3OH$.

Döge[76] observed changes of the width of a band in methyl iodide in different solvents and found a dependence on the polarity of the solvent. This explanation, however, is not sufficient for the effects observed for CCl_4.

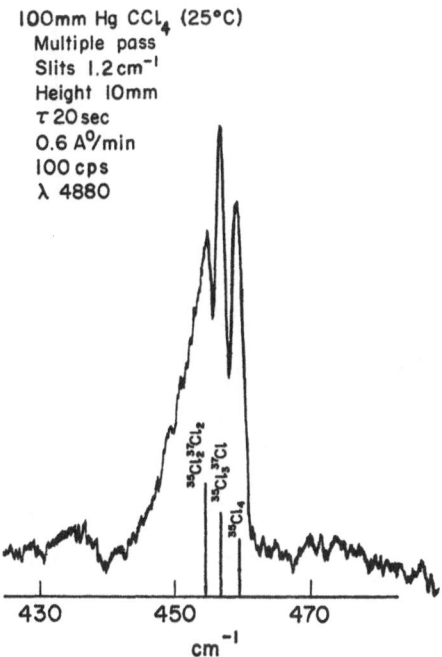

100mm Hg CCl$_4$ (25°C)
Multiple pass
Slits 1.2 cm^{-1}
Height 10mm
τ 20 sec
0.6 A°/min
100 cps
λ 4880

^{35}Cl$_2$^{37}Cl$_2$
^{35}Cl$_3$^{37}Cl
^{35}Cl$_4$

430   450   470
cm^{-1}

Fig. 21. Isotopic structure of the 459
cm^{-1} band of CCl$_4$ in the vapor phase.

Additional profile changes in the wings of the CCl$_4$ band can be recognized in C$_6$H$_{12}$ and CS$_2$. The wings are assigned[77] to hot bands in Fermi resonance with v_1. This assignment is confirmed by the low-temperature spectrum.[78] Figure 23 shows a scan of this band at 77°K made by an instrument manufacturer. An analysis of the profile of the v_1 band in liquid CCl$_4$ at room temperature[24] is shown in Fig. 24.[79]

METHODS AND RESULTS FOR CRYSTALS

A great number of results of investigations of crystals has been reported since the introduction of the laser in Raman spectroscopy. In the last part of this article a short selection is presented with examples of observations of optical phonons, polaritons, magnons, plasmons, and molecular ions in doped crystals.

The chemical implications of single crystal Raman spectroscopy have been outlined by Beattie and Gilson.[80] The basic theory was

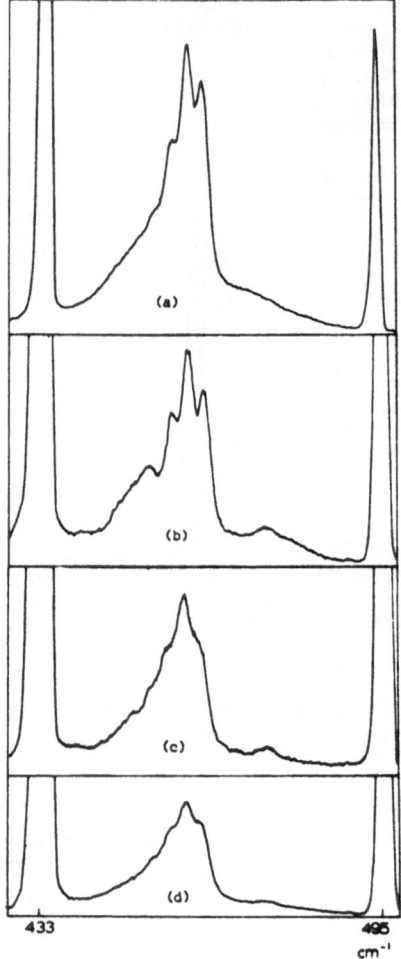

Fig. 22. Effect of solvent on the isotopic
structure of the v_1 Raman line of CCl_4.
a) Pure liquid CCl_4; b) CCl_4 in cyclohexane
1:2; c) CCl_4 in acetone 1:2; d) CCl_4 in
methanol 1:2.[74]

developed by Huang,[111] Poulet,[112] Merten,[113] and Loudon[114] and
was summarized in the classic article by the latter author.[81] For
molecular crystals in which the intermolecular forces are much smaller
than the intramolecular forces one can distinguish between internal
and external vibrations. The internal vibrations correspond to the

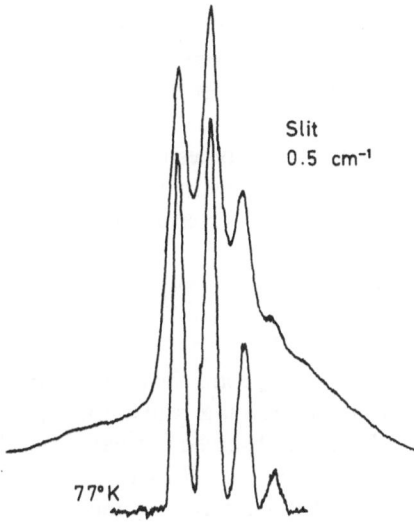

Fig. 23. Low-temperature scan of CCl_4 ν_1 band, recorded with Coderg PH 1 Raman spectrometer ($s = 0.5\,\text{cm}^{-1}$).

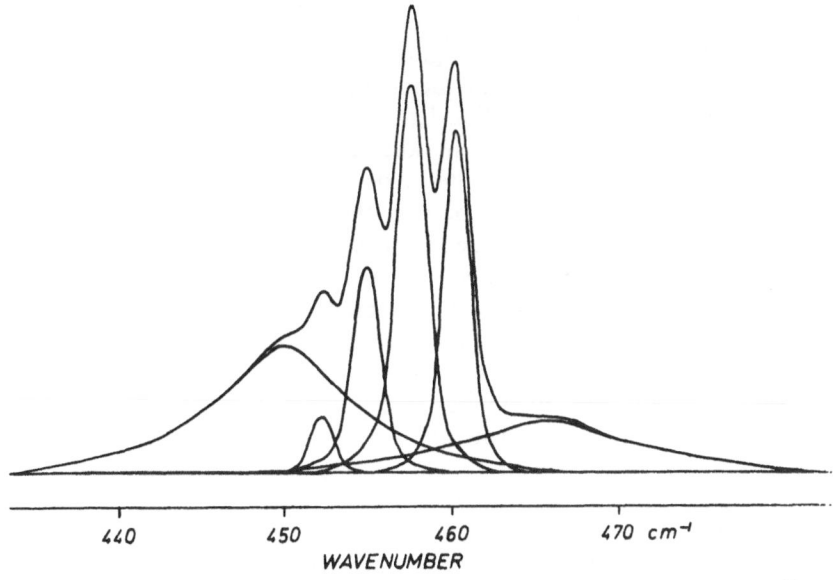

Fig. 24. Analysis of CCl_4 ν_1 band at room temperature with Du Pont curve resolver.[79]

vibrations of the free molecule which are only slightly modified by the crystal field. The external vibrations correspond to translational and rotational vibrations of the molecules in the unit cell. By group theoretical methods the number of vibrations of molecular crystals can be calculated.[82] When the forces between all atoms or ions in the crystal are of the same order, as in alkali halides or diamond, the distinction between internal and external vibrations becomes senseless.

Optical Phonons

For vibrational Raman spectra the derived polarizability tensor is symmetric. For electronic Raman transitions, however, an antisymmetric tensor has also been observed.[2] Apart from this exception it is generally sufficient to record six Raman spectra of a crystal with different orientation and polarization of incident and scattered light with respect to the crystal axes to obtain information on all components of the derived polarizability tensor.

Damen, Porto, and Tell[83] introduced the following notation in their paper on the Raman effect in zinc oxide. With x, y, z as designation for the crystal axes a spectrum is described by four symbols, for example $x(yz)y$. The symbols outside the parenthesis are the direction of incident (left) and scattered light (right), the symbols inside the parenthesis the polarization of the incident (left) and scattered light (right). For the example the exciting light is incident along the x axis and polarized in the y direction, and the scattered light, polarized in the z direction, is observed along the y axis.

A straightforward example is the Raman spectrum of calcite which was studied by Porto and coworkers.[84] Calcite has the symmetry D_{3d} and five Raman-active vibrations are expected: one totally symmetric A_{1g} vibration and two degenerate E_g vibrations of the CO_3^{-2} ion, and two lattice vibrations of species E_g. The A_{1g} vibration should only be observed with the tensor components α_{xx}, α_{yy}, and α_{zz}. With mercury excitation, however, the component α_{xy} also seemed to be observed with considerable intensity. This anomaly even stimulated a number of theoretical papers which sought an explanation. Porto[84] showed, however, that α_{xy} is actually zero. The spectra in Fig. 25 show, that the totally symmetric line 1088 cm^{-1} is only observed with the xx and zz components, while no xy and zx components are detected. The errors in the earlier measurements were caused by too great convergence of the

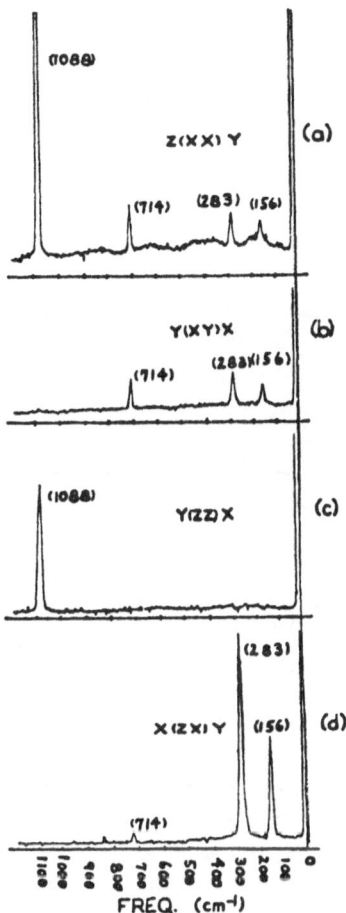

Fig. 25. Raman spectrum of calcite.[84]

exciting light and divergence of the scattered light which made the line appear depolarized. The external vibrations appear strongly with their *zx* components.

The Raman spectrum of α quartz was reinvestigated by Scott and Porto.[85] They showed that four of the eight vibrations of species *E* are split by the electrostatic forces in the crystal into a transverse and a longitudinal component. Figure 26 shows two spectra of a crystal cut at 45° to the *z* axis where the identification of the longitudinal modes is possible.[126]

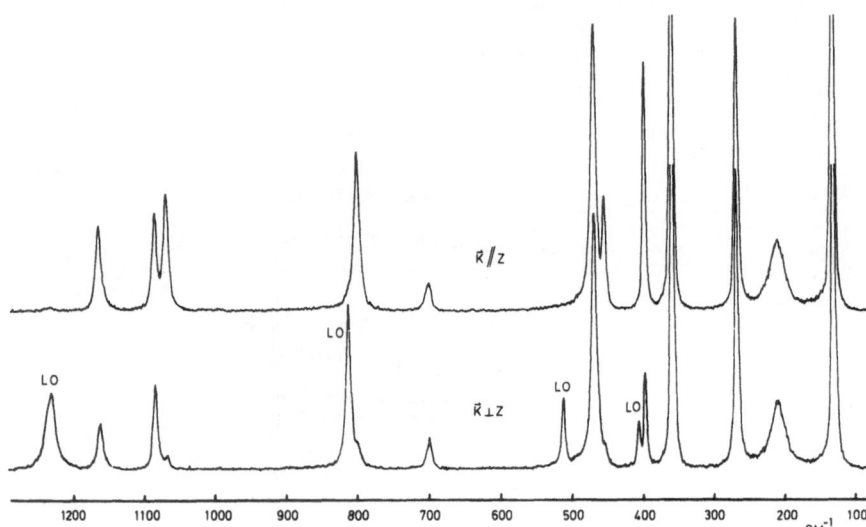

Fig. 26. Identification of *LO* modes in α quartz. (Crystal cut at 45° to z axis.) Scattering plane, xz. Polarization vectors of incident and scattered light $e_i = (0, 1, 0) + (1/\sqrt{2}, 0, -1/\sqrt{2})$ and $e_s = (0, 1, 0) + (1/\sqrt{2}, 0, 1/\sqrt{2})$.[85]

An analogous behavior for a molecular crystal should be observed for benzil, $C_6H_5(CO)_2C_6H_5$. The external spectrum has been investigated by Claus and coworkers.[86] It is shown in Fig. 27 for the orientation $x(zz)y$ for the A_1 modes and $x(yx)y$ and $x(zx)y$ for the E modes. The two latter spectra are different because different tensor elements contribute to the intensity of the lines. The A_1 mode at 30 cm^{-1} is due to a difference vibration because its intensity decreases with temperature, as shown in the upper two spectra. The assignment is summarized in Table 5.

Table 5. Assignment of the External Vibrations of Crystalline Benzil

ν (cm^{-1})	Interpretation
16	E
30	A_1 (hot band)
38	$A_1 + E$
58	E
69	A_1
80	E

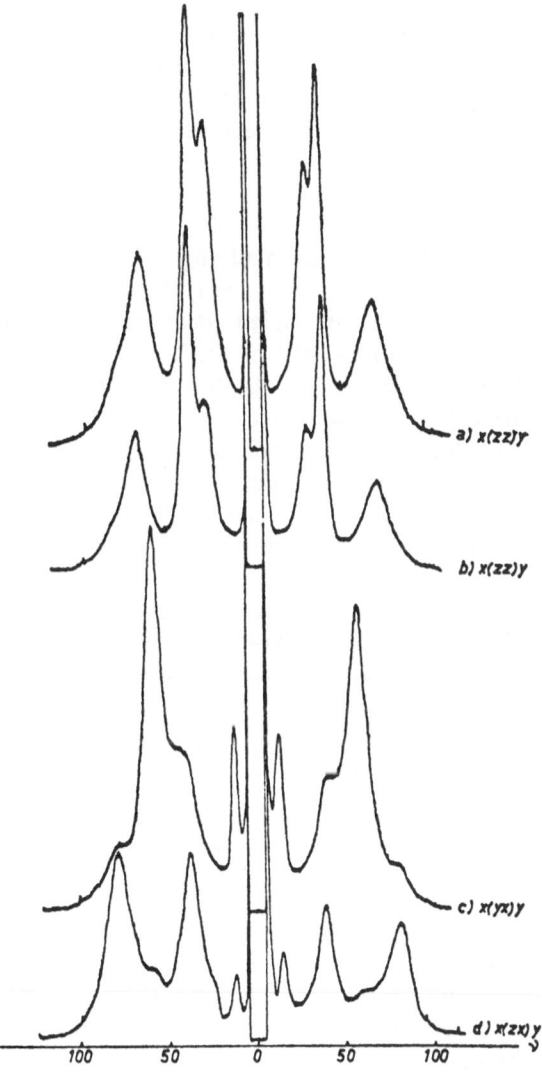

Fig. 27. External vibrations in the Raman spectrum of benzil.
a) A_1 modes at room temperature; b) A_1 modes at $-120°C$;
c) and d) E modes at $-120°C$. Spectral slit width, $2\,cm^{-1}$;
low-temperature cell from Coderg.

The internal spectrum of benzil was studied by Solin and Ramdas[87] who observed and assigned 18 lines. Claus and coworkers[88] were able to detect 88 lines; 35 are assigned to species A_1 and 57 to species E; many broad combination bands were also observed.

For uniaxial crystals with wurtzite structure Arguello and co-workers[115] have demonstrated the frequency shifts of polar phonons for different propagation directions, in agreement with the theory.[112,81]

Polaritons

For forward scattering under small angles the transverse optical phonons can couple with photons with nearly the same wave vector and energy. These mixed states are called polaritons[89] and have been studied by Porto and coworkers[90] in ZnO. Scott, Cheesman, and Porto[91] have also investigated the polariton spectrum of α quartz. In this case eight transverse optical phonons of species E can couple with photons. In Table 6 the wave-number shift of 3 TO lines in α quartz for a change of the scattering angle from 4.8 to 1.2° are given. The wave number of the LO line at 807 cm^{-1} remains unchanged. The agreement between calculated and observed shifts is very satisfactory.

Table 6. Polaritons in α Quartz,[91] Incident Beam, Ordinary, and Extraordinary Scattered Light

Angle	ν (cm^{-1})		
90°	1072	797	450
4.8°	1013	792	443
3.6°	964	785	440
2.4°	918	764	430
1.2°	822	697	407

The polaritons observed in LiNbO$_3$ by the linear[116] and the stimulated[117] Raman effect were used to produce tunable optical emission within[118] and without[119] an external resonator. The corresponding infrared emission at the polariton frequency between 200 and 42 cm^{-1} was also detected.[119]

Another crystal with large nonlinear optical coefficients is LiIO$_3$.[120] The phonon spectrum has been investigated by Claus and coworkers.[121] Polaritons were found to be associated with the transverse optical phonons of species A at 795 and 354 cm^{-1}, and of species

E_1 at 769 cm^{-1}.[121,122] Figure 28 shows the shift of the A polariton for internal scattering angles from 4.8° to 0°.

The behavior of polariton branches at crossover points was investigated theoretically by Lamprecht and Merten[123] and calculated for the example of α quartz.[124] In a more favorable case Claus[125] could directly observe the exchange of intensity between two E polaritons.

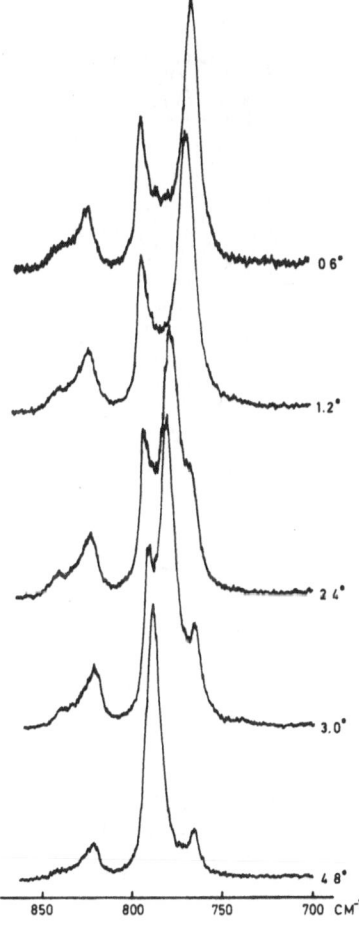

Fig. 28. Polariton associated with A_1 TO mode at 795 cm^{-1} in LiIO$_3$. $x(yy)x$ scattering; incident and ordinary scattered light. Scattering angles given are inside the sample.[121]

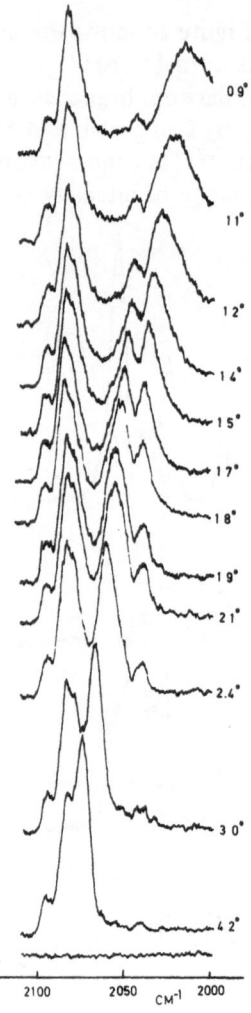

Fig. 29. "Crossing" of polariton branches in uniaxial K_3Cu-$(CN)_4$. $y(zx)y$ scattering. The scattering angles are those inside the sample.

Figure 29 shows the polariton associated with the E phonon at 2082 cm^{-1} moving toward lower wave numbers with decreasing angle, crossing the long-wavelength dispersion curve of a weak E phonon at 2042 cm^{-1}.

Magnons

The Raman spectra of crystals with rutile structure, namely TiO_2, MgF_2, ZnF_2, FeF_2, and MnF_2, which have the symmetry D_{4h} were studied by Porto, Fleury, and Damen.[92] FeF_2 and MnF_2 are of special interest because these crystals become antiferromagnetic at low temperatures. The assignment of the phonon spectra is straightforward.

Below the Neél temperature of FeF_2 (78.5°K) two lines appear at about 50 and 150 cm^{-1} (see Fig. 30).[93] They correspond to the scattering from one and two magnons, namely spin waves, which after observation in antiferromagnetic resonance (for one magnon) and in

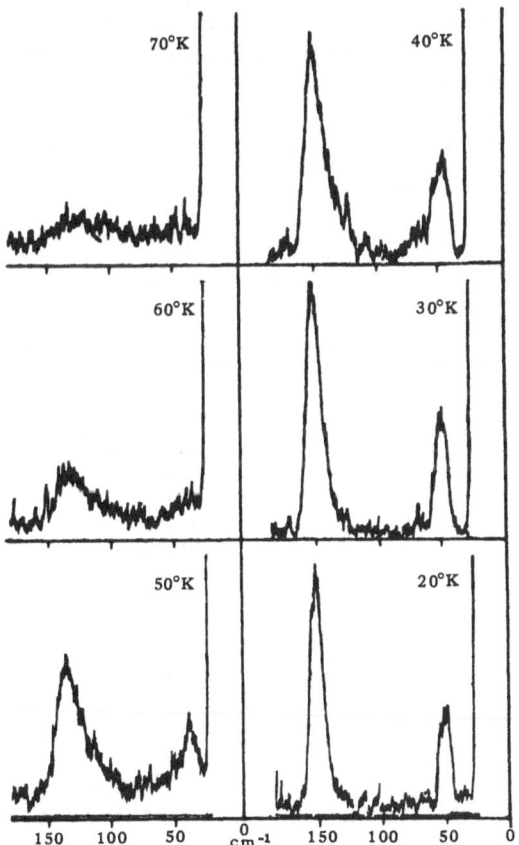

Fig. 30. Raman scattering from magnons in FeF_2. zy component for various temperatures.[93]

infrared absorption (for two magnons) now were also detected by Raman scattering.

Plasmons

Collective vibrations of conduction electrons in semiconductors interact with the longitudinal optical phonons and can be observed in the Raman effect. Mooradian and Wright[10] first observed these plasmon interactions in GaAs with Nd-laser excitation at 1064.8 nm. Figure 31 shows the spectra for various electron concentrations. While the line of the transverse optical phonon remains unchanged at 270 cm^{-1}, the line of the longitudinal optical phonon is shifted and broadened with rising electron concentration. Simultaneously, a new line appears at lower wave numbers and is shifted towards the TO phonon line. The explanation is given in Fig. 32,[94] where the wave-number shifts are plotted as a function of the square root of the electron concentration n. The dashed diagonal corresponds to the calculated plasma frequency according to

$$\omega_p = \left(\frac{4\pi n e^2}{\varepsilon_\infty \cdot m^*} \right)^{\frac{1}{2}}$$

where e is the electron charge, ε_∞ is the optical dielectric constant, and m^* is the conduction-band effective mass. Plasmons and LO phonons

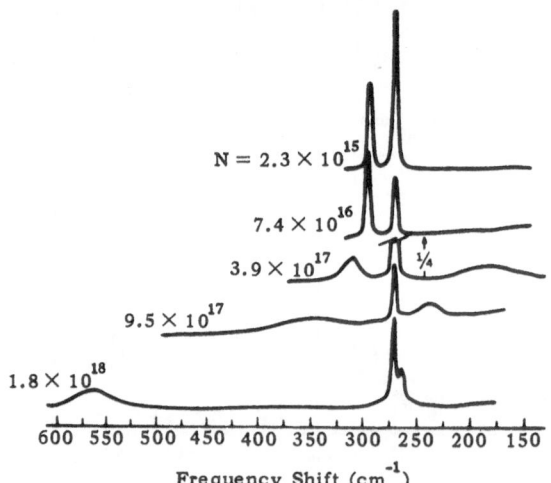

Fig. 31. The anti-Stokes Raman spectrum of n-type GaAs for various carrier concentrations.[9]

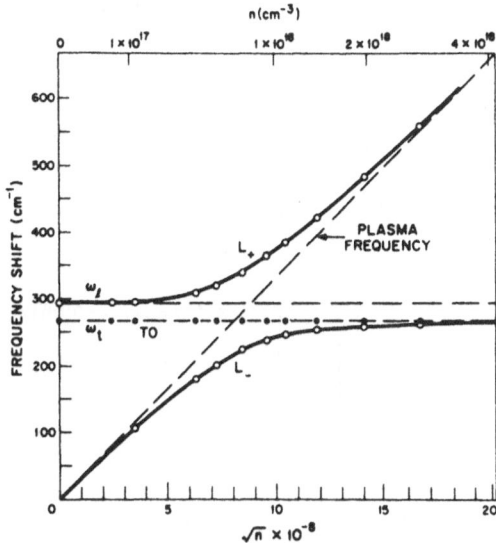

Fig. 32. Frequency shift of the Raman lines in GaAs as function of the square root of the electron concentration.[94]

interact and form L_+ and L_- branches while the TO phonons remain unaffected by the presence of the free carriers. The polarization of the lines is in very good agreement with theoretical expectation.[94]

Mooradian[95] has extended the investigations of plasmons to other semiconductors. Near the exciting line he observed Raman scattering from single electrons corresponding to a Boltzmann distribution at room temperature and a Fermi distribution at low temperatures.

Molecular Ions in Doped Crystals

The new low detection limit of the Raman effect achieved with the argon ion laser makes it possible to study the spectra of molecular ions in crystals at concentrations of the order of 10^{-5}.

Holzer and coworkers[96] studied alkali halide crystals doped with O_2^- ions. Figure 33 shows the vibrational line of the O_2^- ion in KBr. The wave numbers of this line in various crystals are given in the table in Figure 34. In the diagram the wave numbers of O_2^- are plotted against the wave numbers of the vibrational lines of CN^- ions in the

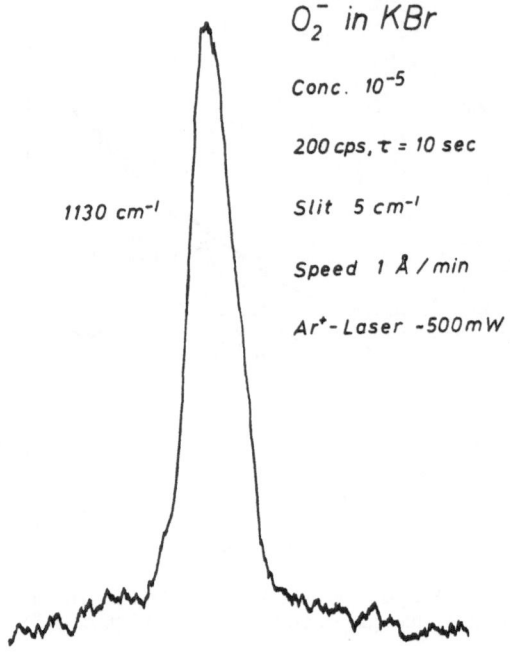

O_2^- in KBr

Conc. 10^{-5}

200 cps, τ = 10 sec

1130 cm^{-1} Slit 5 cm^{-1}

Speed 1 Å / min

Ar$^+$-Laser -500mW

Fig. 33. Raman band of O_2^- in KBr.

same host crystals measured in infrared absorption. The common intersection at 1090 cm^{-1} gives the wave number of the vibration of the free O_2^- ion. The dissociation energy of O_2^- was calculated to 4.1 $\pm$ 0.2 eV.

This study was extended to other negative ions.[97] Raman bands arising from S_2^- and S_3^- were observed in alkali halide crystals which had been heated in the presence of sulfur vapor (Fig. 35). The intensity of the bands is quite high due to a resonance Raman effect. The assignment was confirmed by UV irradiation of the KI crystal after which only the Raman band at 594 cm^{-1} is observed, and only the absorption band at 400 nm corresponding to S_2^- remains unchanged.

The absorption band of the S_3^- species at 610 nm gives rise to a rigorous resonance Raman effect (see Behringer[65]) with considerably increased intensity of the overtones. Figure 36 shows the spectrum of S_3^- in NaCl to the third overtone recorded with constant instrumental conditions.

The sulfur impurity in ultramarine was identified to be S_3^- by its Raman spectrum,[97] which shows a band at 546 cm^{-1} and overtones at

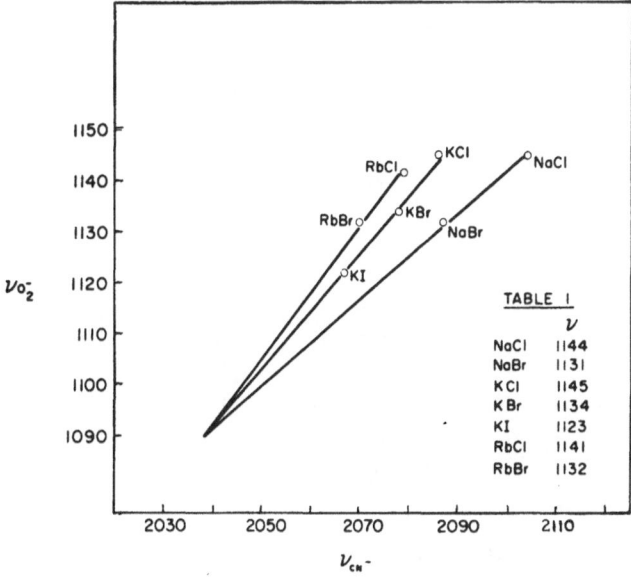

Fig. 34. The Raman wave numbers of O_2^- vs the infrared wave numbers of CN^- in the same crystals (in cm^{-1}).[96]

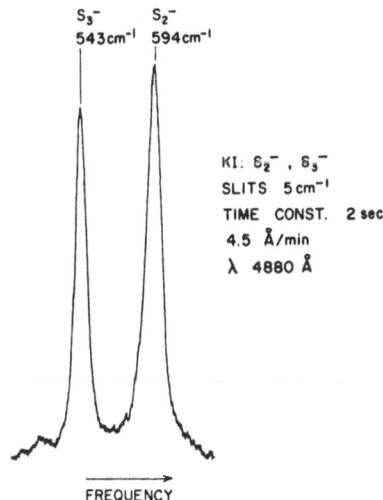

Fig. 35. The Raman bands of S_3^- and S_2^- in a single crystal of KI.

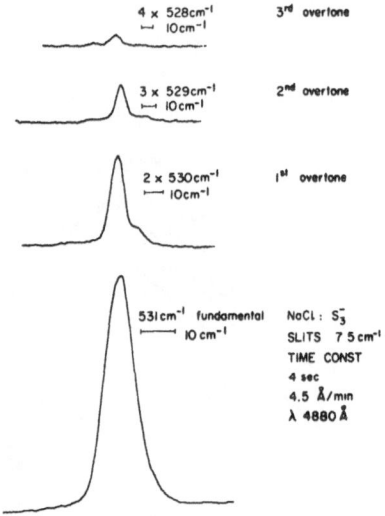

Fig. 36. Resonance Raman effect of S_3^- in NaCl.

2×547, 3×548, and 4×549 cm^{-1} similar to the spectrum shown in Fig. 36. The Raman bands of Se_2^-, SeS^-, and N_2^- in some alkali halide crystals[97] and of NO_2^- and CN^- in KBr[98] were also observed.

CONCLUSIONS

Many of the examples treated in this chapter demonstrate that the use of lasers in Raman spectroscopy has made many new experiments possible which could not have been performed with excitation by discharge lamps. The rapid development of the field in the last few years justifies the expectation that many new results of great interest to chemists and physicists will be obtained in the future.

The development of experimental techniques is far advanced but not yet completed. With new lasers new applications will be possible. With frequency doubling of neodymium, ruby, and argon lasers new exciting frequencies become available. Raman experiments with CO_2 lasers are being performed.

The investigation of adsorbed molecules[99] will gain increasing importance. Intensity measurements in gases will lead to a better understanding of the theory of the Raman effect. Investigations of the dependence of the intensity on the exciting wavelength over a wide range will allow comparison of the results with the theories of the resonance Raman effect.[65,66,100] The observation of Raman spectra of electronically excited molecules seems possible.

The main part of this chapter was finished in March, 1969. I am grateful for the opportunity to make some additions in January, 1970, before the manuscript went into print.

REFERENCES

1. J. A. Koningstein, in: H. A. Szymanski, *Raman Spectroscopy*, Plenum Press, New York (1967), Vol. 1, p. 82.
2. G. B. Wright, Ed., *Light Scattering Spectra of Solids*, Springer, New York (1969).
3. N. Bloembergen, *Am. J. Phys.* **35**: 989 (1967).
4. H. W. Schrötter, *Naturwiss.* **54**: 607 (1967).
5. C. J. Schuler, "Laser Induced Spontaneous and Stimulated Raman Scattering," in: *Progress in Nuclear Energy IX*, Pergamon Press (1968), Vol. 8, p. 2.
6. J. Brandmüller, *Naturwiss.* **54**: 293 (1967).
7. R. E. Hester, *Anal. Chem.* **40**: 320R (1968); R. N. Jones and M. K. Jones, *Anal. Chem.* **38**: 393R (1966); Ya. S. Bobovich, *Usp. Fiz. Nauk* **97**: 37 (1969); P. J. Hendra and P. M. Stratton, *Chem. Reviews* **69**: 325 (1969).
8. B. Schrader, *Chemie-Ingenieur-Technik* **39**: 1008 (1967).
9. A. Mooradian and G. B. Wright, *Phys. Rev. Letters* **16**: 999 (1966).
10. D. Röss, *IEEE J. Quant. Elect.* **2**: 208 (1966).
11. J. Brandmüller, K. Burchardi, H. Hacker, and H. W. Schrötter, *Z. Angew. Phys.* **23**: 112 (1967).
12. M. Delhaye, *Appl. Opt.* **7**: 2195 (1968); *Molecular Spectroscopy*, The Institute of Petroleum, London (1968), p. 275.
13. M. Delhaye and M. Migeon, *Compt. Rend.* **262**: 702, 1513 (1966).
14. G. B. Benedek and K. Fritsch, *Phys. Rev.* **149**: 647 (1966).
15. J. J. Barrett and N. I. Adams, III, *J. Opt. Soc. Am.* **58**: 311 (1968).
16. R. C. C. Leite, S. P. S. Porto, and T. C. Damen, *Appl. Phys. Letters* **10**: 100 (1967).
17. N. J. Bridge and A. D. Buckingham, *Proc. Roy Soc. (London)* **295A**: 334 (1966).
18. J. Topp, H. W. Schrötter, H. Hacker, and J. Brandmüller, *Rev. Sci. Instr.* **40**: 1164 (1969).
19. J. Sharpe, EMI Document, Ref. No. R/P021 (1966).
20. D. O. Landon and S. P. S. Porto, *Appl. Opt.* **4**: 762 (1965).
21. J. A. Koningstein and O. S. Mortensen, *J. Mol. Spectry.* **27**: 343 (1968).
22. C. R. Vidal, *Z. Instrumentenk.* **74**: 316 (1966).
23. W. D. Gunther, E. F. Erickson, and G. R. Grant, *Appl. Opt.* **4**: 512 (1965); G. R. Grant and W. D. Gunther, *Rev. Sci. Instr.* **36**: 1511 (1965); J. R. Sizelove and J. A. Love, III, *Appl. Opt.* **6**: 443 (1967); H. Hora and R. Kantlehner, *Solid State Comm.* **4**: 557 (1966); T. Hirschfeld, *Appl. Opt.* **7**: 443 (1968); J. B. Oke and R. E. Schild, *Appl. Opt.* **7**: 617 (1968).
24. J. Brandmüller, K. Burchardi, H. Hacker and H. W. Schrötter, *Z. Angew Phys.* **22**: 177 (1967).
25. W. C. Elmore, *Nucleonics* **6**: 26 (1950).
26. Y.-H. Pao and J. E. Griffiths, *J. Chem. Phys.* **46**: 1671 (1967).
27. R. R. Alfano and N. Ockman, *J. Opt. Soc. Am.* **58**: 90 (1968).

28. V. P. Kozlov, *Opt. Spectry.* **24**: 453 (1968).
29. W. Kiefer and H. W. Schrötter, *Z. Angew. Phys.* **25**: 236 (1968).
30. E. Ziegler and E. G. Hoffmann, *Österr. Chem. Z.* **68**: 319 (1967).
31. J. R. Ferraro, in: H. A. Szymanski, *Raman Spectroscopy*, Plenum Press, New York (1967), Vol. 1, p. 44.
32. *Perkin–Elmer Instrument News* **19**(2): 5 (1968).
33. H. S. Haber, H. J. Sloane, and R. C. Hawes, Symposium on Molecular Structure and Spectroscopy, Columbus, Ohio, 1967.
34. M. Schubert, *Expt. Techn. Physik* **8**: 155 (1960).
35. B. Schrader, F. Nerdel, and G. Kresze, *Z. Physik. Chem. (N.F.)* **12**: 132 (1957); *Z. Anal. Chem.* **170**: 43 (1959).
36. J. Brandmüller, H. Hacker, W. Kiefer, H. W. Schrötter, J. Topp, and M. Wahl, *Verhandl. DPG (VI)* **4**: 388 (1969).
37. H. W. Schrötter, *Z. Angew. Phys.* **12**: 275 (1960).
38. G. R. Harrison, *MIT Wavelength Tables*, John Wiley and Sons, New York (1952).
39. A. Opler, *J. Opt. Soc. Am.* **41**: 349 (1951).
40. G. Strey, *Spectrochim. Acta* **25A**: 163 (1969).
41. K. Burchardi, Diplomarbeit Universität München, 1967 (unpublished).
42. E. B. Wilson, J. C. Decius, and P. C. Cross, *Molecular Vibrations*, McGraw-Hill, New York (1955).
43. B. Kellerer, H. Hacker, and J. Brandmüller (work in progress).
44. L. A. Woodward, in: H. A. Szymanski, *Raman Spectroscopy*, Plenum Press, New York (1967), Vol. 1, p. 1.
45. A. Weber, S. P. S. Porto, L. E. Cheesman, and J. J. Barrett, *J. Opt. Soc. Am.* **57**: 19 (1967).
46a. A. Lau, *Expt. Techn. Physik* **16**: 253 (1968).
46b. H. H. Claassen, H. Selig, and J. Shamir, *J. Appl. Spectry.* **23**: 8 (1969).
47. C. Allemand (private communication).
48. T. C. Damen, R. C. C. Leite, and S. P. S. Porto, *Phys. Rev. Letters* **14**: 9 (1965); S. P. S. Porto, *J. Opt. Soc. Am.* **56**: 1585 (1966).
49. J. G. Skinner and W. G. Nilsen, *J. Opt. Soc. Am.* **58**: 113 (1968).
50. W. R. Hess, H. Hacker, H. W. Schrötter, and J. Brandmüller, *Z. Angew. Phys.* **27**: 233 (1969).
51. G. W. Chantry, *Spectrochim. Acta* **21**: 1007 (1965).
52. W. F. Murphy, W. Holzer, and H. J. Bernstein, Symposium on Molecular Structure and Spectroscopy, Columbus, Ohio, September 1968.
53. W. F. Murphy, M. V. Evans, and P. Bender, *J. Chem. Phys.* **47**: 1836 (1967).
54. G. E. Walrafen, *J. Chem. Phys.* **46**: 1870 (1967).
55. M. M. Sushchinskii and Z. M. Muldakhmetov, *Opt. Spectry.* **16**: 128 (1964).
56. R. D. Mair and D. F. Hornig, *J. Chem. Phys.* **17**: 1236 (1949).
57. H. J. Bernstein and G. Allen, *J. Opt. Soc. Am.* **45**: 237 (1955).
58. D. G. Rea, *J. Opt. Soc. Am.* **49**: 90 (1959).
59. Yu. I. Naberukhin, *Opt. Spectry.* **13**: 278 (1962).
60. F. J. McClung and D. Weiner, *J. Opt. Soc. Am.* **54**: 641 (1964); D. Weiner, S. E. Schwarz, and F. J. McClung, *J. Appl. Phys.* **36**: 2395 (1965).
61. G. Bret, *Compt. Rend.* **260**: 6323 (1965).
62. G. Eckhardt and W. G. Wagner, *J. Mol. Spectry.* **19**: 407 (1966).
63. W. D. Johnston, Jr., I. P. Kaminow, and J. G. Bergman, Jr., *Appl. Phys. Letters* **13**: 190 (1968).
64. H. H. Hacker, Dissertation Universität München, 1968 (unpublished).
65. J. Behringer, in: H. A. Szymanski, *Raman Spectroscopy*, Plenum Press, New York (1967), Vol. 1, p. 168.

66. H. Buyken, K. Klauss, and H. Moser, *Ber. Bunsenges. Physik. Chem.* **71**: 578 (1967).
67. W. F. Murphy, W. Holzer, and H. J. Bernstein, *J. Appl. Spectry.* **23**: 211 (1969).
68. J. J. Barrett, 9th European Congress on Molecular Spectroscopy, Madrid, September 1967.
69. H. J. Bernstein, 9th European Congress on Molecular Spectroscopy, Madrid, September 1967; W. F. Murphy and H. J. Bernstein, Symposium on Molecular Structure and Spectroscopy, Columbus, Ohio, September 1967.
70. W. Holzer and H. Moser, *J. Mol. Spectry.* **13**: 430 (1964).
71. W. Holzer, W. F. Murphy, and H. J. Bernstein, *J. Chem. Phys.* **52**: 399, 469 (1970).
72. W. Clements and B. Stoicheff, *Appl. Phys. Letters* **12**: 246 (1968).
73. M. Scotto, *J. Chem. Phys.* **49**: 5362 (1968).
74. H. W. Kroto and Y. H. Pao, *J. Opt. Soc. Am.* **58**: 479 (1968).
75. W. F. Murphy, H. J. Bernstein, and H. W. Schrötter (unpublished).
76. G. Döge, *Z. Naturforsch.* **23a**: 1405 (1968).
77. H. Horiuti, *Z. Physik* **84**: 380 (1933).
78. H. J. Bernstein and W. F. Murphy (private communication).
79. H. H. Hacker and H. W. Schrötter, *J. Chem. Phys.* **49**: 3325 (1968).
80. I. R. Beattie and T. R. Gilson, *Proc. Roy. Soc. (London)* **A307**: 407 (1968).
81. R. Loudon, *Adv. Physics* **13**: 423 (1964); **14**: 621 (1965).
82. S. S. Mitra, *Solid State Physics: Advances in Research and Applications*, Vol. 13, New York (1962); B. Schrader, Habilitationsschrift, Universität Münster, 1968; S. Bhagavantam and T. Venkatarayudu, *Theory of Groups and Its Application to Physical Problems*, Academic Press, New York–London (1969).
83. T. C. Damen, S. P. S. Porto, and B. Tell, *Phys. Rev.* **142**: 570 (1966).
84. S. P. S. Porto, J. A. Giordmaine, and T. C. Damen, *Phys. Rev.* **147**: 608 (1966).
85. J. F. Scott and S. P. S. Porto, *Phys. Rev.* **161**: 903 (1967).
86. R. Claus, H. H. Hacker, H. W. Schrötter, J. Brandmüller, and S. Haussühl, *Phys. Rev.* **187**: 1128 (1969).
87. S. A. Solin and A. K. Ramdas, *Phys. Rev.* **174**: 1069 (1968).
88. R. Claus, H. W. Schrötter, J. Brandmüller, and S. Haussühl, *J. Chem. Phys.* **52** (in press).
89. J. J. Hopfield, *Phys. Rev.* **112**: 1555 (1958); C. H. Henry and J. J. Hopfield, *Phys. Rev. Letters* **15**: 964 (1965).
90. S. P. S. Porto, B. Tell, and T. C. Damen, *Phys. Rev. Letters* **16**: 450 (1966).
91. J. F. Scott, L. E. Cheesman, and S. P. S. Porto, *Phys. Rev.* **162**: 834 (1967).
92. S. P. S. Porto, P. A. Fleury, and T. C. Damen, *Phys. Rev.* **154**: 522 (1967).
93. P. A. Fleury, S. P. S. Porto, L. E. Cheesman, and H. J. Guggenheim, *Phys. Rev. Letters* **17**: 84 (1966).
94. A. Mooradian and A. L. McWhorter, *Phys. Rev. Letters* **19**: 849 (1967).
95. A. Mooradian, *Verhandl. DPG (VI)* **4**: 127 (1969); in: O. Madelung, *Festkörperprobleme IX, Advances in Solid State Physics*, Pergamon-Vieweg, Braunschweig (1969), p. 74.
96. W. Holzer, W. F. Murphy, H. J. Bernstein, and J. Rolfe, *J. Mol. Spectry.* **26**: 534 (1968); J. Rolfe, W. Holzer, W. F. Murphy, and H. J. Bernstein, *J. Chem. Phys.* **49**: 963 (1968).
97. W. Holzer, W. F. Murphy, and H. J. Bernstein, *J. Mol. Spectry.* **32**: 13 (1969).
98. R. Callender and P. S. Pershan, *Phys. Rev. Letters* **23**: 947 (1969).
99. P. H. Hendra, in: *Molecular Spectroscopy*, The Institute of Petroleum, London (1968), p. 285.
100. A. C. Albrecht and J. Tang, in: H. A. Szymanski, *Raman Spectroscopy*, Plenum Press, New York (1970), Vol. 2, p. 33.
101. G. Michel, *Spectrochim. Acta* **25A**: 517 (1969).

102. H. W. Schrötter and J. Bofilias, *J. Mol. Structure* **3**: 242 (1969).
103. M. M. Sushchinskii, *Spektry Kombinatsionnogo Rasseyaniya Molekul i Kristallov. Fizika i Tekhnika Spektralnogo Analiza*, Izdatelstvo "Nauka," Moscow (1969).
104. J. A. Topp and W. Schmid (to be published).
105. J. R. Scherer, Symposium on Molecular Structure and Spectroscopy, Columbus, Ohio, September 1968; International Conference on Raman Spectroscopy, Ottawa, Canada, August 1969.
106. W. F. Murphy, Symposium on Molecular Structure and Spectroscopy, Columbus, Ohio, September 1969.
107. J. Bofilias, Dissertation Universität München (unpublished).
108. J. J. Barrett and A. Weber, *J. Opt. Soc. Am.* **59**: 1531A (1969); **60**: 70 (1970).
109. J. Behringer, *Z. Physik* **229**: 209 (1969).
110. W. Kiefer, International Conference on Raman Spectroscopy, Ottawa, Canada, August 1969.
111. K. Huang, *Proc. Roy. Soc. (London)* **A208**: 352 (1951).
112. H. Poulet, *Ann. Phys. (Paris)* **10**: 908 (1955).
113. L. Merten, *Z. Naturforsch.* **15a**: 47 (1960); **17a**: 65 (1962).
114. R. Loudon, *Proc. Phys. Soc. (London)* **82**: 393 (1963); *Proc. Roy. Soc. (London)* **A275**: 218 (1963).
115. C. A. Arguello, D. L. Rousseau, and S. P. S. Porto, *Phys. Rev.* **181**: 1351 (1969).
116. H. E. Puthoff, R. H. Pantell, B. G. Huth, and M. A. Chacon, *J. Appl. Phys.* **39**: 2144 (1968).
117. S. K. Kurtz and J. A. Giordmaine, *Phys. Rev. Letters* **22**: 192 (1969).
118. J. Gelbwachs, R. H. Pantell, H. E. Puthoff, and J. M. Yarborough, *Appl. Phys. Letters* **14**: 258 (1969).
119. J. M. Yarborough, S. S. Sussman, H. E. Puthoff, R. H. Pantell, and B. C. Johnson, *Appl. Phys. Letters* **15**: 102 (1969).
120. G. Nath and S. Haussühl, *Appl. Phys. Letters* **14**: 154 (1969); *Phys. Letters* **29A**: 91 (1969).
121. R. Claus, H. W. Schrötter, H. H. Hacker, and S. Haussühl, *Z. Naturforsch.* **24a**: 1733 (1969).
122. R. Claus, *Z. Naturforsch.* **25a**: 306 (1970).
123. G. Lamprecht and L. Merten, *Phys. Stat. Sol.* **35**: 353 (1969).
124. L. Merten, *Z. Naturforsch.* **24a**: 1878 (1969).
125. R. Claus, *Phys. Letters* **31A**: 299 (1970).
126. R. Claus, Dissertation, Universität München, 1970 (unpublished).
127. R. L. Schwiesow, *J. Opt. Soc. Am.* **59**: 1285 (1969).

Chapter 4

Low-Frequency Raman Spectra of Liquids*

L. A. Blatz

University of California
Los Alamos Scientific Laboratory
Los Alamos, New Mexico

INTRODUCTION

Since 1964 it has been possible to observe,[1,2] clearly, low-frequency (less than 100 to 150 cm^{-1}) Raman lines from liquids. Previous observations[3-5] of this low-frequency Raman spectral region from liquids were, in general, subject to disagreements, both theoretical and experimental.[4,5]

If a conventional Raman spectrophotometer is used to observe a spectral Rayleigh line scattered first from air (or a low-pressure gas) and then from various liquids, it will be noted that the lines from liquids are larger and have varied half-widths (line width at half-maximum intensity), have different shapes, and have tails called "Rayleigh line wings" extending often to 100 to 150 cm^{-1}. This phenomenon was first noted by Raman and Krishnan,[6] and although it has been studied extensively[2-5] for the past 40 years, many questions are still unanswered.

This chapter is concerned mainly with the experimental methods that have been developed to observe these low-frequency Raman lines. The study of these lines has yielded some information on the intermolecular structures of liquids and liquid solutions, and more information can be expected as the experimental methods are refined.

EXPERIMENTAL METHODS

When monochromatic light of wavelength λ_0 is used to illuminate a spectrograph which has gratings, prisms, or combinations of the two as the dispersing elements, it is noticed that stray light can be seen over the entire spectral region and that this stray light is particularly intense

*Work done under the auspices of the U.S. Atomic Energy Commission.

in the immediate neighborhood of λ_0, where it will interfere with the observation of any relatively weak spectral lines that may be present.

Stray light arises from many sources,[7-10] and various methods have been devised to reduce it; but no way has been found to eliminate it completely without, at the same time, almost entirely eliminating any weak lines in the immediate neighborhood of intense lines.

Ideally, the low-frequency Raman spectrum of a liquid can be observed best by the illumination of a volume of a 100% pure, dust-free, nonfluorescent liquid (or a solution made from pure components) with intense, parallel, monochromatic (perhaps polarized) exciting light followed by complete elimination of the exciting light before it leaves this volume. In addition the sample container should not fluoresce, contribute its own Raman light in the region of interest, or cause any exciting light to leave the sample in the same direction as the Raman light which is to be detected.

This chapter will consider various experimental methods that have been proposed in attempts to approximate the ideal situation. The methods will be taken up in the order that the hardware occurs in a conventional Raman spectrophotometer, from light source and sample to the detection system.

All but one of the methods depend on the reduction of either the ratio R of Rayleigh-to-Raman line intensities, or the relative electronic response to these intensities. The attenuation function is defined as the fractional reduction in each line intensity, at each wavelength or frequency, over the region of interest, produced by some method. If this function can be determined accurately, it should be possible to correct each intensity ordinate at each frequency of the observed spectrum. Experimentally, no accurate attenuation function for even one Raman spectrophotometer and attenuation method has been published to date. The "Intensity decrease" column of Table 1 in Blatz[2] is one example of an attenuation function which, although probably adequate for determining line frequencies to the stated precision, was very tedious to apply, and, because of its increasingly rapid change, became increasingly inaccurate below about 40 cm^{-1}. This attenuation function does show that when the Rayleigh line ordinate was decreased 48-fold more than a Raman line ordinate at 50 cm^{-1}, low-frequency Raman lines from liquids were clearly observed.

A continuous light source, even if it has a constant and known intensity distribution, probably cannot be used to obtain an accurate attenuation function, because the intensity ordinate at any given wavelength is composed of intensities from many other wavelengths which change in complex and uncertain ways during a determination of the attenuation function.

Purification of Samples

Liquids can be purified by successive distillation, preferably without ebullition, into a cold trap. If the liquids or solutions are not used within a short time, they should be stored in inert containers.

Solutions of salts in water can be purified by treatment with a pure grade of adsorbent carbon.[11,12] The author has found that treatment with a pure carbon,[11] Spheron-6, increased the transmission of a very good grade of conductivity water by 5% at 2536 Å for 10-cm cells.

In all cases the solutions should be filtered finally through the smallest pore-size filters available, such as the Millipore or Gelman membrane filters, and some of the liquid or solution should be used to rinse the previously cleaned sample container.

These purification procedures have yielded as much as a 40-fold[12] reduction of continuous background in the low-frequency region near 4358 Å, and several-fold reductions of continuous background are commonly achieved.

Sample Container

The design and treatment of sample containers have been discussed elsewhere.[2,13–15] It is especially important to minimize R at this point before other methods are used following the sample container.

It has been found[2] that when circular, cylindrical sample containers were painted black, properly, and arranged so that a small air gap existed between the tube and the outer part of the Raman cell, up to 10-fold reductions in the 4358-Å Rayleigh line intensities were recorded with only small changes in low-frequency Raman line intensities.

Light Sources

The subject of light sources has been considered in numerous papers. See, for example, those by Ferraro[15] and Koningstein.[16]

Some sources, such as Toronto-type mercury arcs, that are operated at too high electric currents and temperatures, yield excessive amounts of continuous background.

Light Filters

Various light filters, or their equivalent, placed usually, but not always, immediately after the sample container, will reduce the ratio R of the light leaving the filter.

Mercury-Vapor Filter. One of the oldest methods[17] is the use of the resonance 2536-Å mercury line as the exciting line followed by its absorption in a mercury-vapor filter placed immediately after the sample container. This method is of limited applicability since (1) most liquids have large light absorbances at 2536 Å, (2) impurities in water solutions of salts and acids give rise to fluorescence and large light absorbances which are difficult to reproduce from one sample to another or for one sample at different times, and (3) quartz optics fluoresce on illumination with 2536-Å light. By the use of this method a line from water at about 60 cm^{-1} was observed (Hibben,[17] pp. 320 and 327). However, samples can be cleaned adequately[11,12] and non-fluorescent quartz is now available for use as sample, light source, and mercury-vapor filter containers.

Light Absorption Filters. Water solutions of potassium ferricyanide have been used to decrease the ratio R beyond 4358 Å, and various other absorption filters have also been used (cf. Hibben,[17] pp. 29 and 30). Although none of these filters produces a sufficiently large decrease in R to permit the clear observation of low-frequency Raman lines from liquids, some may be useful in combination with other methods.

Double Monochromators. An exceptionally efficient filter is obtained by the use of another dispersing monochromator to precede the dispersing instrument. Commercially available, two-grating, double monochromators combined with careful use of items mentioned above enabled Walrafen[18] and Gasner (Blatz,[2] p. 844) to detect Raman light at about 60 cm^{-1} from liquid water.

The commercially available, two-grating, double monochromators used in Raman spectrophotometers have not been adequate[2] to observe clearly the low-frequency lines listed in Table 2 of Blatz.[2] The benzene 75-cm^{-1} line, in Table 2 of Blatz,[2] has less than 1% of the intensity of the parent 4358-Å Rayleigh scattered line, on the basis of line heights. As the Rayleigh line frequency is approached (during a scan with two-grating, double monochromators), more and more of the Rayleigh line and stray light are transmitted to the second monochromator (especially with the wider slits necessary to avoid undue intensity losses) where they can give rise to stray light, especially intense in the neighborhood of the Rayleigh line.

"Random errors in the groove spacing of gratings give rise to stray light in the neighborhood of a spectral line. These random errors are greater in gratings ruled by the interferometrically controlled method than in those ruled by traditional methods."[19] Probably diffuse (non-specular) reflection at the surfaces of the other optical parts of grating instruments contributes to this stray light in the neighborhood of the Rayleigh line and perhaps multiple reflections (cf. also Sawyer,[8] pp. 189 and 190, and Barnes,[7] pp. 169 and 170) and fluorescence also contribute.

Interference Filters. The author was able to considerably extend the observation of the low-frequency region around 4358 Å just by the use of several different interference filters, and the results were better the less the value of R at a given frequency in this region. One interference filter[20] could be set so as to weaken the 4358-Å line more than 10-fold while transmitting greater than 70% of the region above 100 cm^{-1}. It seems probable that two of these filters in series, arranged to give less than 1% transmission at the exciting line frequency and 25 to 50% transmission at 100 cm^{-1} (cf. Blatz,[2] Table 1), would enable Raman spectrophotometers to observe some low-frequency Raman lines.

Some of the lines listed in Blatz,[2] Table 2 were observed by McDevitt[21] by the use of an interference filter which transmitted about 0.1% of the exciting line and about 5% at 50 cm^{-1}.

It would, of course, be very useful if a narrow-band-blocking interference filter were available which would decrease the transmission of the exciting line more than 100-fold and transmit more than 50% of the low-frequency region.

Some disadvantages of these interference filters are (1) they are temperature sensitive, (2) the three filters used by the author deteriorated with age so that the steep increase in transmission beyond 4358 Å became much less steep, (3) the very rapid decrease of transmission near the exciting line makes the observation of some Raman lines below about 50 cm^{-1} difficult or impossible, and makes the accurate correction of observed line profiles in the 0- to 50-cm^{-1} region very difficult, and (4) the filter has different transmission characteristics for each angle of incident light and, therefore, the use of a parallel beam of light needed for optimum performance would be achieved with a large Raman intensity loss.

Narrow Pass Interference Filters. Since interference filters reflect the light they do not transmit, narrow band-pass filters can be used to transmit the exciting line away from (and reflect the Raman light toward) the dispersing instrument.[22] Interference filters are available from various manufacturers with peak transmissions of about 50% at the laser line wavelengths, λ_0 = 4880, 5145, or 6328 Å, and with transmissions of 10% at $\lambda_0 \pm$ about 10 Å. After one transmission and reflection, R is decreased about two-fold at ± 10 Å or more from λ_0. Although this is far from the 25- or 50-fold attenuation that is needed, it is a worthwhile result. Greater attenuation probably can be achieved by more than one[22] transmission and reflection, although the Raman radiation will be weakened and diffused by the multiple reflections and the narrow band pass of the filter (or filters) will be partly lost as a result of the various angles of incident light necessitated by the more complicated optical arrangement.

A more convenient arrangement could be obtained by two or more narrow band-blocking filters set in series. Two blocking filters, each with 50% transmission at λ_0 and 90% at $\lambda_0 \pm 10$ Å, would decrease R about four-fold when used in series, and the narrow band could be more easily maintained by the use of parallel incident light.

Interference-Filter Light Chopper. Narrow band interference filters with either a maximum (pass filter) or minimum (blocking filter) transmission at the exciting line wavelength λ_0, can be used in another way.[23] A circular disc, composed half of the interference filter and half of a neutral density filter with the same light transmission at λ_0, is arranged to interrupt periodically the light beam before it enters the dispersing instrument. The Raman light is chopped while the exciting line is not. The Raman light signal is then amplified with a suitable frequency sensitive lock-in amplifier. Figures 1–4 show some of the low-frequency Raman spectra of liquids[24] obtained in this way with a narrow pass interference filter with about 50% transmission at λ_0 = 6328 Å (80 mW, He–Ne, laser source) and nearly zero at ± 50 cm^{-1}, and a Spex model 1400, two-grating, double monochromator.

Some disadvantages of this system, in addition to those previously mentioned in connection with interference filters, include the fact that (1) the use of a narrow pass interference filter with a 50% transmission at λ_0 produces about a four-fold loss of Raman line intensity (most of this intensity will be recovered if blocking filters are available and can be used) and that (2) it is difficult to make filters with precisely the same

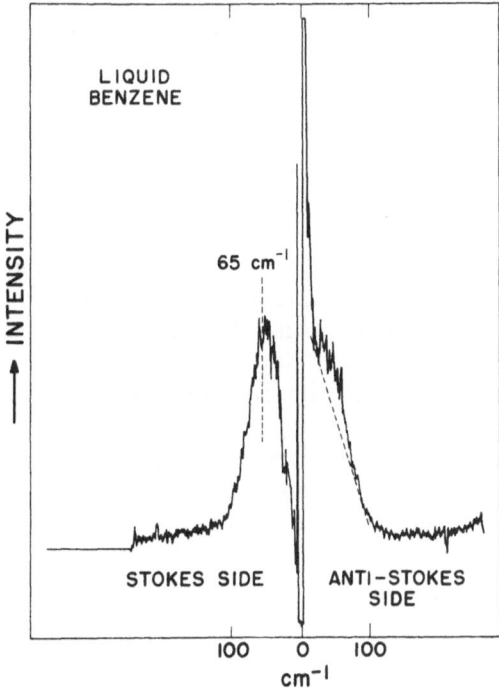

Fig. 1. The low-frequency spectrum of liquid benzene.[24] The spectrum was taken with a lock-in amplifier, and the filter–chopper was tilted to favor the Stokes side. Scan, 100 cm^{-1}/min; time constant, 1 sec; slits, 5 cm^{-1}.

transmission over their entire area, and the varying light transmission of the incident beam, as the filters revolve, will give rise to excess noise. Variations in the transmission of the exciting line are especially serious because of its relatively much greater intensity.

Polarized Light Techniques

Since the low-frequency Raman lines (observed to date) are depolarized while the Rayleigh line is at least partly polarized (in the usual Raman spectrophotometric applications), the use of a polarizer following the sample cell will decrease R. This technique, combined with a theoretical study of the Rayleigh line and its wing, was used by Starunov[25] and Zaitsev[26] to observe low-frequency Raman lines from some liquids. This technique, combined with light-scattering measurements,[14,27,28] has been used to study the low-frequency Raman region (cf. also Craddock, Jackson, and Powles[29] and Rank, Hollinger, and Eastman[30]).

Increased Angular Dispersion

If the angular dispersion of a Steinheil three glass prism spectrograph is increased,[1] at 4358 Å, to more than three times the value at the minimum deviation setting of the prisms, large increases in the ratio of Raman line to continuous background heights for liquids are obtained in the low-frequency region. Figure 5 shows the low-frequency spectrum of water obtained in this way. The spectrum is very nearly the same[2,18] as that obtained with two different, two-grating, double monochromators.

It is postulated that, with increased angular dispersion, the stray light in the immediate neighborhood of an intense spectral line remains close to the parent line where it does not interfere as much with a weak

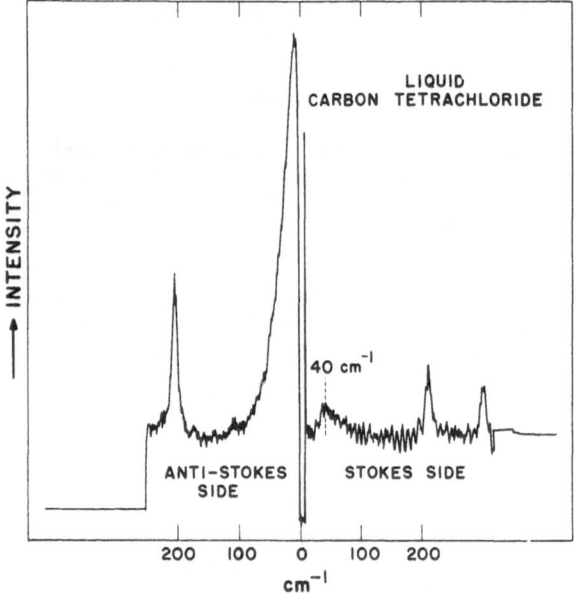

Fig. 2. The low-frequency spectrum of liquid carbon tetrachloride.[24] The spectrum was taken with a lock-in amplifier, and the filter–chopper was tilted to favor the Stokes side. Scan, 100 cm^{-1}/min; time constant, 1 sec; slits, 5 cm^{-1}.

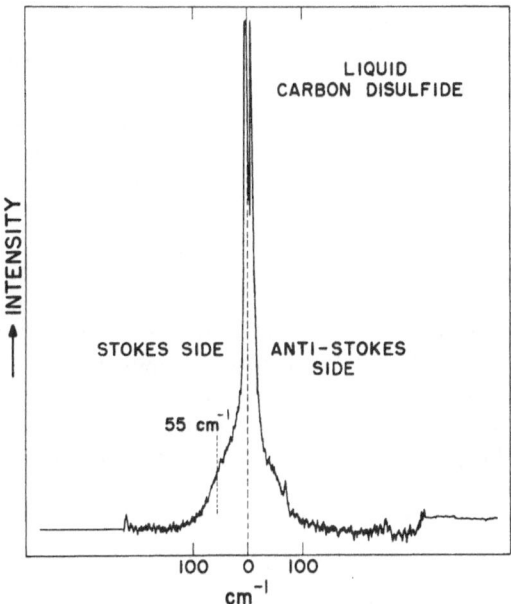

Fig. 3. The low-frequency spectrum of liquid carbon disulfide.[24] The spectrum was taken with a lock-in amplifier,
and the filter–chopper was set to pass the Stokes and anti-
Stokes sides symmetrically. Scan $100 \, cm^{-1}$/min; time
constant, 1 sec; slits, $5 \, cm^{-1}$.

line which is now farther removed (in space), in the focal plane. The
following observations are consistent with this postulate: (1) the
continuous background below about $1000 \, cm^{-1}$, as recorded by a
Raman spectrophotometer,[1,2] remains almost flat, with increased
angular dispersion, to frequencies much closer to the exciting line, as
compared to the lesser angular dispersion case (cf. Barnes,[7] p. 170); (2)
visual observation indicates that the 4358-Å stray light is still very
intense in the immediate neighborhood of 4358 Å and considerably
less intense in other regions; (3) the author was able to obtain 4- to
10-fold increases of Raman line to continuous background heights in
one series of (unpublished) observations on various liquids in the low-
frequency region. However, the increased angular dispersion did not[2]
permit the clear observation of the benzene 75-cm^{-1} line.

Blumenfeld and Fast[31] used a double-monochromator arrangement combined with a large dispersion spectrograph to observe a low-
frequency line in liquid formic acid.

Some disadvantages of the use of large increases in angular dispersion with prism instruments are that (1) the greater curvatures of the spectral lines are more difficult to match with the curved exit slits, (2) there is some loss of intensity in the prisms especially as the exit angle of refraction approaches the critical angle, and (3) the greater the dispersion the more sensitive the system becomes to temperature fluctuations, and mechanical vibrations and instabilities.

Mechanical Barriers

After the light has been dispersed, barriers (baffles) can be placed in the light beam to decrease R before the light reaches the detector. Figure 6 shows one way this may be achieved.

Hibben[17] (cf. Zirnit and Sushchinskii[32]) placed a baffle in front of the photographic plate, which prevented the exciting line from striking

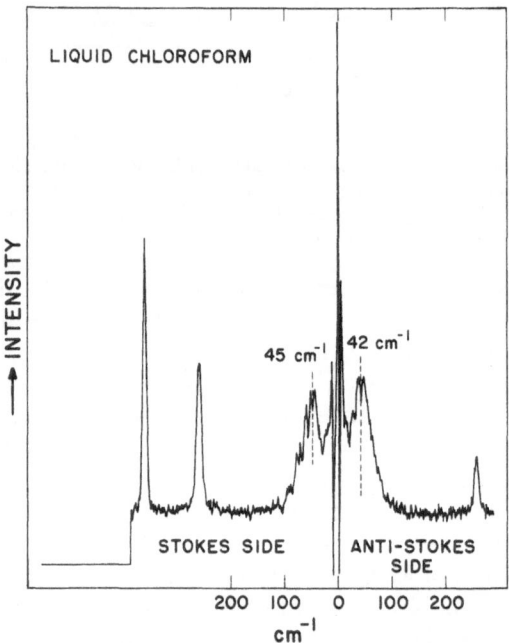

Fig. 4. The low-frequency spectrum of liquid chloroform.[24] The spectrum was taken with a lock-in amplifier, and the filter-chopper was set to pass the Stokes and anti-Stokes sides symmetrically. Scan $100 \, cm^{-1}$/min; time constant, 1 sec; Slits, $5 \, cm^{-1}$.

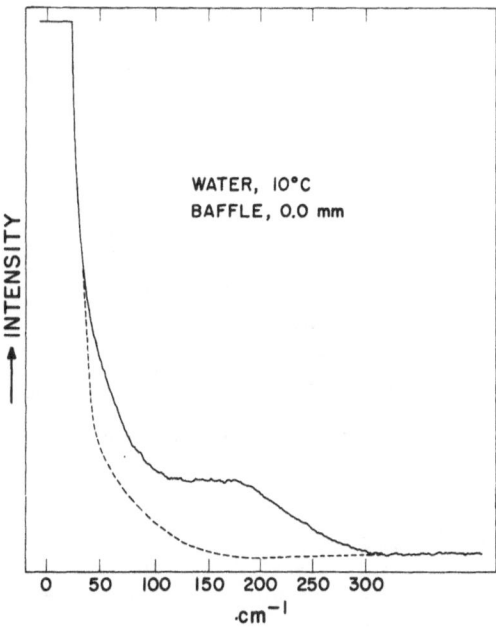

Fig. 5. The low-frequency spectrum of water at 10°C. The spectrum was taken with a Steinheil, three glass prism spectrograph with an angular dispersion (at 4360 Å) which had been increased[1] to 3.3 times the value at the minimum deviation setting of the prisms. The dotted line is the background line. Scan, 19.7 cm^{-1}/min; time constant, 10 sec; slits, 4.9 cm^{-1}.

the plate, reduced the halation, and made possible somewhat clearer observation of the low-frequency region.

The author[1,2] placed a baffle just before the camera lens (Fig. 7) of a Steinheil three glass prism spectrograph. The circular beam which left the collimator lens overlapped the sharp, straight, refracting edge of the first prism so that the blue light beam (4358 Å) had the shape of a semi-ellipse in the plane just in front of the camera lens. This device, which made possible large reductions in R, prevented almost all of the exciting frequency from entering the two-element, four air–glass surface, camera lens. The method is perfectly general and should also be useful for the observation of weak lines (less than 1% as intense) in the neighborhood of any intense lines.

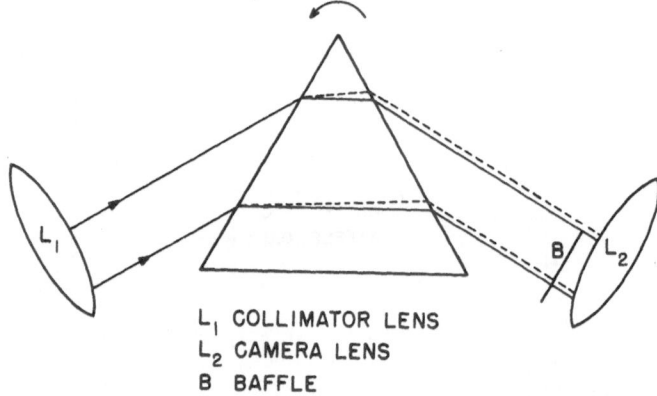

L₁ COLLIMATOR LENS
L₂ CAMERA LENS
B BAFFLE

Fig. 6. A simplified illustration of how a blackened baffle may be used to absorb the very intense 4358-Å light (solid line) while a fraction of the light at 4400 Å (dashed line) is permitted to pass through the camera lens to the detector. The prism is rotated (arrow) in the direction shown to decrease the angle of incidence and increase the angular dispersion.

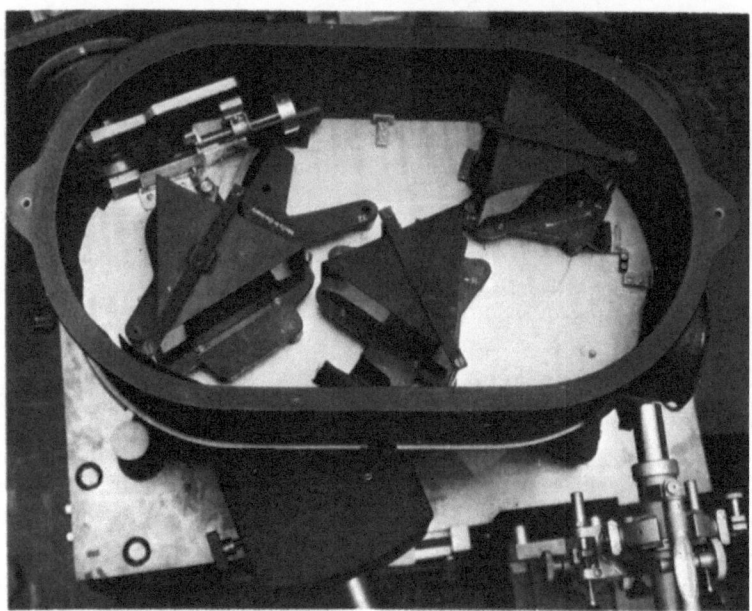

Fig. 7. A photograph of the baffle mechanism (upper left) that was used to translate the blackened plate in a plane just in front of the camera lens of a Steinheil three glass prism spectrograph. The prisms have all been rotated counterclockwise to decrease the angles of incidence and increase the angular dispersion.

Figures 8–11 show the low-frequency spectra of various liquids obtained[1,2] by this technique combined with increased angular dispersion of the spectrograph.

The use of the baffle technique[1] produces considerable losses in Raman line intensities, and these losses are greater the lower the line frequency (cf. Blatz,[2] Table 1, Column 3). However, the chief disadvantage of the baffle technique is the very abrupt change in R below 30 to 40 cm^{-1} which makes the accurate determination of an attenuation function, and therefore the accurate correction of Raman line ordinates, difficult. The chief advantage of this technique is that it has permitted the observation of weaker and lower frequency Raman lines than any other method except, perhaps, the filter–chopper method of

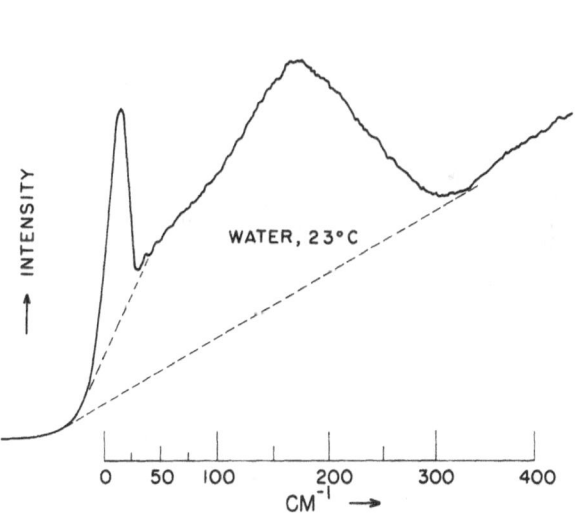

Fig. 8. The low-frequency spectrum of water at 23°C taken with increased angular dispersion and the baffle set to decrease the Rayleigh scattered 4358-Å line ordinate at 0 cm^{-1} more than 1000-fold (the unbaffled or residual Rayleigh line is near 0 cm^{-1}) while the Raman line ordinate at 175 cm^{-1} was decreased less than two-fold. The low-frequency line at about 60 cm^{-1} is clearly visible. The dotted lines show how the background was drawn (cf. Fig. 5). Scan, 19.7 cm^{-1}/min; time constant, 10 sec; slits, 4.9 cm^{-1}.

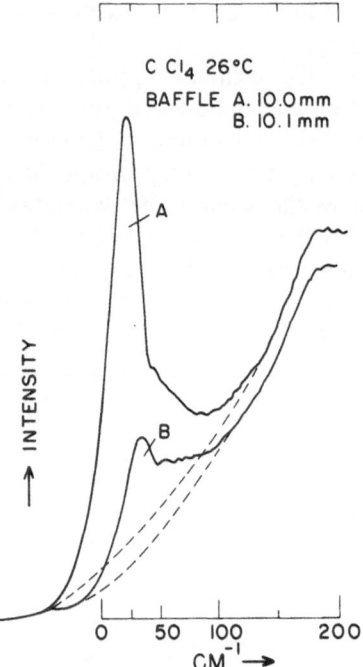

Fig. 9. The low-frequency spectrum of liquid carbon tetrachloride.[2] The height of the residual Rayleigh scattered 4358-Å line was decreased 3.5-fold from A to B by a 0.1-mm motion of the baffle. The dotted lines are background lines. Scan, 19.7 cm^{-1}/min; time constant, 10 sec; slits, 4.9 cm^{-1}.

Landon,[23,24] previously discussed, and also the next method to be taken up.

Electronic Method

The intensity scale can be compressed so as to amplify the low-frequency Raman lines considerably more than the Rayleigh line peak region. A logarithmic slide wire on a strip-chart recorder or an amplifier with a logarithmic response can be used to plot the logarithm of the intensity *vs* frequency. Figure 12[24] shows the spectrum of liquid benzene and Fig. 13[33] shows the spectrum of liquid carbon tetrachloride obtained by recording the logarithm of intensity as a function of frequency. The chief disadvantages of this method are the complexities involved in drawing background lines and then in determining frequencies, shapes, and relative intensities of the Raman lines. However, the line from liquid carbon tetrachloride is one of the most difficult of the low-frequency Raman lines to observe, and, as Fig. 13 shows, it is clearly observable by this method. It seems that it would be desirable to develop this method further.

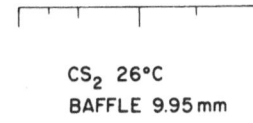

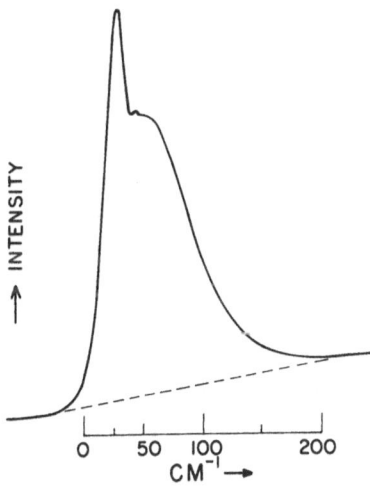

Fig. 10. The low-frequency spectrum of liquid carbon disulfide.[2] The Rayleigh scattered 4358-Å line ordinate was decreased 2000- to 3000-fold to leave the residual sharp peak near 25 cm^{-1} while the Raman radiation ordinate at 50 cm^{-1} has been decreased 25-fold. The Raman line is clearly asymmetric.[2] Scan, 19.7 $\text{cm}^{-1}/\text{min}$; time constant, 5 sec; slits, 4.9 cm^{-1}.

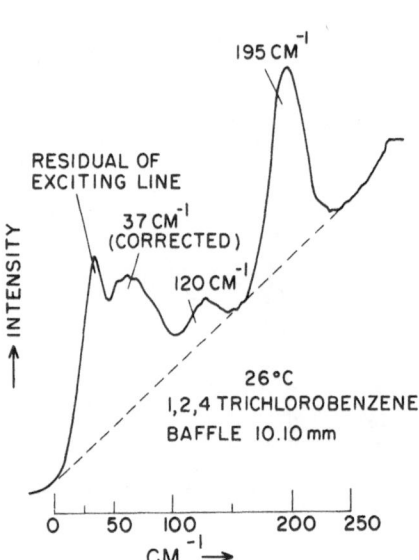

Fig. 11. The low-frequency spectrum of liquid
1,2,4-trichlorobenzene.[2] The Raman line at
37 cm^{-1} clearly shows a peak intensity.
Although the 37-cm^{-1} line seems to be asymmetric, this remains to be established after
further improvements in technique. Scan,
$19.7 \text{ cm}^{-1}/\text{min}$; time constant, 5 sec; slits,
4.9 cm^{-1}.

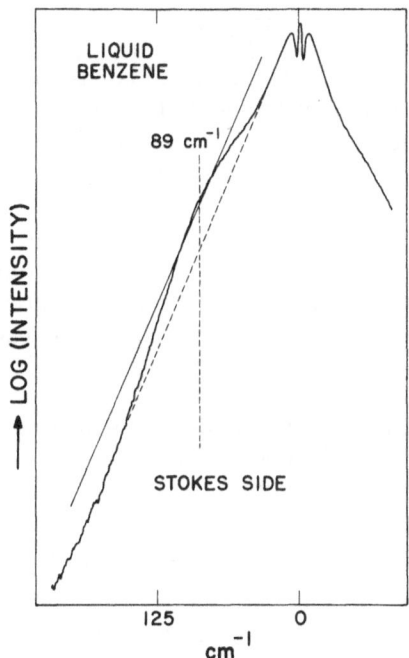

Fig. 12. The low-frequency spectrum of liquid benzene[24] taken with an electronic system which recorded log intensity as a function of frequency. Scan, 50 cm^{-1}/min; time constant, variable; slits, 2 cm^{-1}.

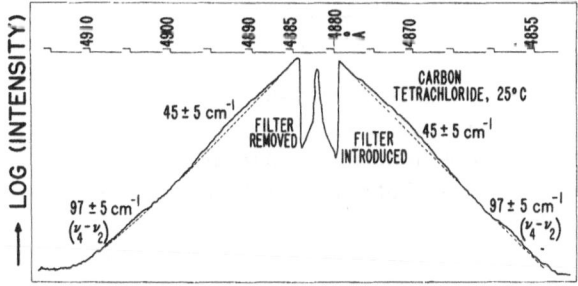

Fig. 13. The low-frequency spectrum of liquid carbon tetrachloride[33] taken with an electronic system which recorded log intensity as a function of frequency. Both the Stokes and anti-Stokes regions near the argon ion laser, 4880-Å exciting line are shown. The line at 97 ± 5 cm^{-1} is a difference line. The dotted lines are the background lines.

RESULTS

Perhaps as many as several hundred liquids and liquid solutions have been found to give low-frequency Raman lines, although a large part of these data are in papers now in process of publication. Only one liquid, n-hexadecane, failed[2] to yield a low-frequency Raman line (to the author's knowledge).

Table 1 lists the frequencies of some low-frequency Raman lines from liquids, all at room temperature, obtained by various methods. In view of the difficulties associated with these observations, the numbers obtained by various methods are in reasonably good agreement at this time.

Table 1. Frequencies in cm^{-1} of Low-Frequency Raman Lines from Liquids at Room Temperature

	Benzene	Toluene	Carbon tetrachloride	Carbon disulfide	Chloroform	Formic acid	Acetic acid
Blatz[2] and Waldstein[34]	75 ± 5	71 ± 5	30 ± 10	35 ± 10	40 ± 10	71 ± 5	47 ± 7
Landon[24]							
Filter–chopper	89		42	54	48		
Electronic method	67		40	55	44		
McDevitt[21]	71						
Starunov and Zaitsev[25,26]	80	80		71	60		50
Bernstein[33]	88		45 ± 5				
Blumenfeld and Fast[31]						82	

The intensity profiles of the low-frequency region from formic and acetic acids[34] were corrected only to the extent necessary to obtain the line frequencies, that is, in the region of the peaks; therefore, the relative intensities for formic acid obtained by Blumenfeld and Fast[31] may not be in disagreement.

There is some doubt as to whether the intensity of the low-frequency Raman radiation in carbon tetrachloride and carbon disulfide (and perhaps some other liquids such as n-heptane[2]) has a zero, intermediate, or maximum value at zero cm^{-1} (cf. Blatz,[2] Kastha,[4] Gabelnick and Strauss,[14,35] Starunov and Zaitsev,[25,26] Shapiro and Broida,[27] McClintock,[28] and Venkateswarlu and Thyagarajan[34]). For carbon tetrachloride the results of Bernstein[33] (Fig. 13) and Landon[24]

(Fig. 2) show definite intensity peaks, while the frequency published by the author[2] was obtained by assuming a peak, that is, that the intensity is zero[4,36] (or at least is not a maximum) at zero cm^{-1}. The data of Gabelnick and Strauss[14] (Fig. 3) and McClintock[28] (Fig. 4–7) show maximum intensities at zero cm^{-1}.

The spectrum of 1,2,4-trichlorobenzene[2] (Fig. 11) clearly shows a peak at 37 cm^{-1}, but the spectra of carbon tetrachloride[2] (Fig. 9) and carbon disulfide[2] (Fig. 10) (cf. Shapiro and Broida[27]) do not so clearly exhibit peak intensities at the values 30 and 35 cm^{-1}, respectively (Table 1).

Types of low-energy vibrations which can or do give rise to low-frequency Raman and far infrared lines from liquids have been listed by various authors.[2,5,12,14,21,25–28,34,37–39]

The low-energy vibrational lines have been compared to results obtained from inelastic scatter of neutrons from liquids.[2,40,41]

FUTURE

The chief difficulty with all existing methods, at this time, is the lack of accurate attenuation functions. However, improvements in the methods of detecting low-frequency Raman lines are certain to be made in the near future, and these improvements will undoubtedly result in the removal of the (perhaps apparent) disagreements at this time.

Undoubtedly, more information on the structure of liquids will be obtained from further studies of their low-frequency Raman lines.

REFERENCES

1. L. A. Blatz, *Spectrochim. Acta* **21**: 1973 (1965).
2. L. A. Blatz, *J. Chem. Phys.* **47**: 841 (1967).
3. G. Herzberg, *Infra-Red and Raman Spectra of Polyatomic Molecules*, Van Nostrand, New York (1945), pp. 531–537.
4. G. S. Kastha, *Indian J. Phys.* **32**: 473 (1958), and references therein.
5. I. L. Fabelinskii, *Molecular Scattering of Light*, Plenum Press, New York (1968), Chapt. 8, p. 101, and references therein.
6. C. V. Raman and K. S. Krishnan, *Nature* **122**: 278, 882 (1928).
7. B. T. Barnes, "The Spectrometer as an Optical Instrument," in: W. E. Forsythe, *Measurement of Radiant Energy*, McGraw-Hill, New York (1937).
8. R. A. Sawyer, *Experimental Spectroscopy*, 2nd ed., Dover Publications, New York (1963), pp. 121, 180, and following pages.
9. *The Spex Speaker* **11**(1): (1966); **11**:(2): 4 (1966); **12**(4): (1967).
10. J. V. Kline and D. W. Steinhaus, *Appl. Opt.* **7**: 2015 (1968).
11. L. A. Blatz, *Anal. Chem.* **33**: 249 (1961).
12. L. A. Blatz and P. Waldstein, *J. Phys. Chem.* **72**: 2614 (1968).

13. T. T. Wali and D. F. Hornig, *J. Chem. Phys.* **45**: 3424 (1966).
14. H. S. Gabelnick and H. L. Strauss, *J. Chem. Phys.* **49**: 2334 (1968).
15. J. R. Ferraro, "Advances in Raman Instrumentation and Sampling Techniques," in: H. A. Szymanski, *Raman Spectroscopy*, Plenum Press, New York (1967).
16. J. A. Koningstein, "Laser Raman Spectroscopy," in: H. A. Szymanski, *Raman Spectroscopy*, Plenum Press, New York (1967).
17. J. H. Hibben, *The Raman Effect and Its Chemical Applications*, Reinhold, New York (1939).
18. G. E. Walrafen, *J. Chem. Phys.* **44**: 1546 (1966).
19. D. W. Steinhaus, Los Alamos Scientific Laboratory, private communication, December 1968 (cf. references 9, 10).
20. This filter was made by Thin Film Products, Inc., Cambridge, Mass., in June, 1964. The manufacture of this filter was apparently quite difficult, but probably the same or even better filters can be made now (January 1969).
21. N. T. McDevitt, Seventh National Meeting, Society for Applied Spectroscopy, Chicago, Ill., May 14, 1968.
22. B. Schrader and W. Meier, *Z. Naturforsch.* **21a**: 480 (1966).
23. D. O. Landon, Seventh National Meeting, Society for Applied Spectroscopy, Chicago, Ill., May 15, 1968.
24. D. O. Landon, Spex Industries Inc., Metuchen, N.J., private communication, May 1968.
25. V. S. Starunov, *Opt. Spectry.* **18**: 165 (1965); V. S. Starunov, "Study of the Spectrum of the Thermal and Stimulated Molecular Scattering of Light in Liquids," in: D. V. Skobel'tsyn, *Optical Studies in Liquids and Solids*, Vol. 39, Plenum Press, New York (1969).
26. G. I. Zaitsev and V. S. Starunov, *Opt. Spectry.* **19**: 497 (1965); **22**: 221 (1967).
27. S. L. Shapiro and H. P. Broida, *Phys. Rev.* **154**: 129 (1967).
28. M. McClintock, doctoral thesis, University of Colorado, 1967.
29. H. C. Craddock, D. A. Jackson, and J. G. Powles, *Mol. Phys.* **14**: 373 (1968).
30. D. H. Rank, A. Hollinger, and D. P. Eastman, *J. Opt. Soc. Am.* **56**: 1057 (1966).
31. S. M. Blumenfeld and H. Fast, *Spectrochim. Acta* **24A**: 1449 (1968).
32. U. A. Zirnit and M. M. Sushchinskii, *Opt. Spectry.* **16**: 489 (1964); U. A. Zirnit, "Study of the Rotational Oscillation of Molecules in Liquids by the Raman Method," in: *Optical Studies in Liquids and Solids*, cited in reference 25 above.
33. H. J. Bernstein, National Research Council of Canada, Ottawa, Canada, private communication, December, 1968.
34. P. Waldstein and L. A. Blatz, *J. Phys. Chem.* **71**: 2271 (1967).
35. H. L. Strauss, "Vibrational Spectroscopy," in: *Ann. Rev. Phys. Chem.* **19**: 419 (1968), p. 437.
36. K. Venkateswarlu and G. Thyagarajan, *Z. Physik.* **154**: 81 (1959).
37. W. J. Hurley, I. D. Kuntz, and G. E. Leroi, *J. Am. Chem. Soc.* **88**: 3199 (1966).
38. J. W. Brasch, Y. Mikawa, and R. J. Jakobsen, *Appl. Spectry. Rev.* **1**(2): 187 (1968); *Appl. Spectry.* **22**: 641 (1968).
39. H. J. Clase and H. W. Kroto, *Mol. Phys.* **15**: 167 (1968).
40. S. Trevino, *Appl. Spectry.* **22**: 659 (1968).
41. J. J. Rush, *J. Chem. Phys.* **47**: 3936 (1967).

Chapter 5

High- and Low-Temperature Raman Spectroscopy

Ronald E. Hester

University of York
York, England

INTRODUCTION

A huge majority of the papers published on Raman spectroscopy have been concerned only with spectra obtained from samples held at the arbitrary temperature of the spectrometer sample housing. This usually has been just a few degrees above room temperature, due to some slight heating effects of the light source, which most commonly has been a powerful mercury arc lamp. The reason for this state of affairs can in part be ascribed to the simple laziness of practical spectroscopists, who have generally found that there are many systems of interest which can be made to yield worthwhile results from the most unsophisticated experimental setups, and who have been content to leave the design problems of spectroscopy to the manufacturers of commercial instruments. However, in their defence it must be said that the problems of varying and controlling sample temperature over a wide enough range to be useful have, until recently, been quite difficult to solve.

The most common type of Raman sample, a clear liquid, transparent to the exciting frequency, scatters only about $10^{-6}\%$ of the incident light in the Raman effect. Accurate measurement of the resulting low levels of light has generally been a difficult enough business without the further complication brought about by the inevitable attenuation of both the incident and the scattered light by thermostat equipment. The need to retain high-level illumination of samples from almost all directions has in turn severely restricted the possibilities for temperature control. With the conventional Wood's[1] sample-tube design coupled with a mercury lamp of the Toronto-arc[2] type, these problems are further compounded by the awkward shape of the area to be controlled, and by the restricted space available between the tube and the lamp coils for locating control equipment.

Since at the time of writing there still are many laboratories where traditional arc- and discharge-lamp excitation methods are employed, an attempt will be made in this chapter to cover some of the more successful techniques for high- and low-temperature spectroscopy which have been devised specifically for these circumstances. However, it is clear that more promising methods for controlled temperature Raman spectroscopy become available when a laser is used in place of the helical mercury lamp as a light source. The advantages of laser excitation in this context are enormous. Not only can the laser itself be located far away from the sample, thus enabling elaborate and bulky temperature control systems to be constructed, but the highly directional character of this light source obviates the need for more than a small window into the sample container to allow entry of the laser beam, and one to allow exit of the scattered light. The high intensity of laser light, particularly when fully utilized by focussing or multipassing techniques, virtually eliminates the problems previously encountered with low light levels emerging from multiple-windowed cells, such as are used in low-temperature work. A range of cell designs for high- and low-temperature Raman studies using lasers will be presented in the following sections. Most of these have been used with continuous lasers, usually the He–Ne or Ar ion gas lasers, but pulsed lasers could be used equally well with most of them and offer promise for future studies of short-lived species.

HIGH-TEMPERATURE TECHNIQUES

In the first volume of this book Ferraro[3] described and reproduced a drawing of a high-temperature cell assembly designed by Walrafen[4] for use with the Cary 81 spectrometer. Although it was possible to reach temperatures close to 1100°C with this cell, which used the radio-frequency-heating technique, it called for very extensive modifications of the standard Cary instrument, including construction of four new vertical mercury arc lamps which did not couple with the RF-heating coils, and was overelaborate for general adoption. The earlier design by Walrafen, Irish, and Young[5] was, by contrast, very much simpler. Since this simpler arrangement is useful with any standard horizontally mounted arc or discharge lamp, and enabled good quality Raman spectra to be obtained at temperatures up to 620°C, it is reproduced here in Fig. 1. This design was based on still earlier cells used for molten salt work by Janz and his coworkers,[6] and uses a widely

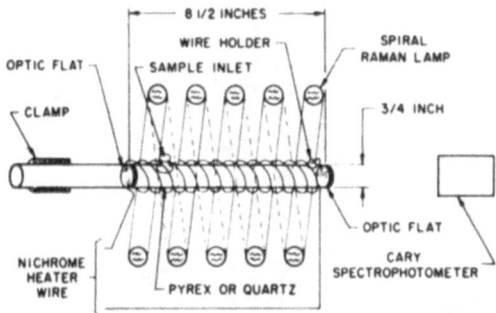

Fig. 1. Schematic diagram of high-temperature Raman assembly.[5]

spaced, wire-wound quartz tube to contain the hot sample. The Janz cell has been described in Volume 1 of this book,[7] as have cells by Young and Westerdahl[8] and Bues,[9] so that no further account of these is necessary here.

Among the earliest applications of lasers to high-temperature Raman studies was that made by Clarke, Solomons, and Balasubrah-manyam,[10] who used a pulsed ruby laser for excitation of spectra from fused salts, although they pointed out the fact that continuous-wave (cw) gas lasers would be still more advantageous. Figures 2 and 3 show the main features of their optical system and sample-furnace assembly.

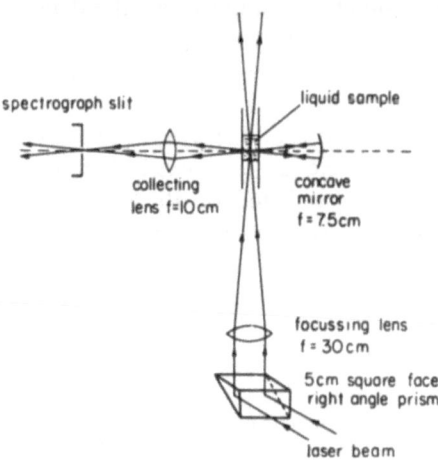

Fig. 2. Optical system for laser excitation of Raman spectra.[10]

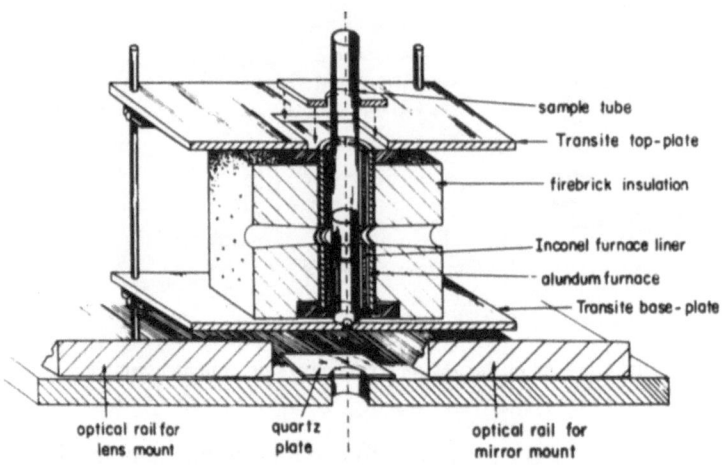

Fig. 3. Section of a sample furnace for high-temperature work.[10]

The focussing of the laser beam produces a very bright image well suited to photoelectric recording techniques (though Clarke *et al.* used photographic recording), and enables small sample sizes to be used. Temperatures in the region of 1000°C were recorded with the assembly shown in Fig. 3, with quartz being used as the cell material. At such temperatures it was found to be necessary to prevent direct light emission from the hot sample from falling continuously on the detector (photographic plate), and this was achieved by use of a shutter which was left closed except for the short periods when the laser was discharged.

A modification of this cell is at present used routinely at York and has been described by Clarke and Hester.[11] This modified cell has been employed with cw, He–Ne gas-laser excitation, and has proved to be fully as convenient as predicted by the earlier authors. The main features of this design are shown in Fig. 4. The furnace was fabricated from a block of silicone-resin-impregnated asbestos (Sindanyo), eight loops of high-current resistance wire being inset into the inner wall of the block and wound parallel to its longitudinal axis to minimize temperature gradients. The furnace power is controlled by a variac up to a maximum of 650 W, producing a maximum sample temperature of about 1200°C, this being measured with a Pt–Pt, 13% Rh thermocouple implanted in the base of the furnace block. Small holes cut in the block allow passage of the focussed laser beam, and larger cone-shaped holes cut at right angles to these allow exit of scattered light. The laser beam is focussed at the center of the sample by a microscope objective lens, and multipassing of the laser beam is achieved by placing a concave mirror below the cell and furnace assembly. This increases by several-fold the intensity of the scattered light, which is collected by a large aperture magnifying lens and focussed on the entrance slit of a double monochromator. Backscattered light is collected by a concave mirror and reflected back through the sample to the monochromator. The sample tube, based on the design previously described,[10] was made in two concentric sections from silica, the inner tube being closed at its lower end by a fine-porosity silica sinter through which a molten example can be forced under pressure of dry nitrogen. Since laser excitation with liquids is particularly sensitive to suspended solid particles this filtration process is an important feature. Overall, this is a very simple cell to make and use. It has the further advantage over the original design[10] of avoiding passage of the laser beam through a liquid meniscus, which can be a source of instability. To avoid interference by the thermal radiation from very hot samples it is a simple matter to chop the laser beam (800 Hz is used at York) before the sample, and to discriminate against the dc generated by the photomultiplier tube through use of phase-sensitive detection techniques. The use of narrow band-pass filters also eliminates most of the ac noise originating in the hot sample.

High-temperature samples which are sufficiently corrosive to attack Pyrex and silica cells are not uncommon and pose special problems in container design. Arighi and Evans[12] have described a cell for use with such corrosive systems at temperatures up to 300°C, and although these authors made use of mercury arc excitation it appears

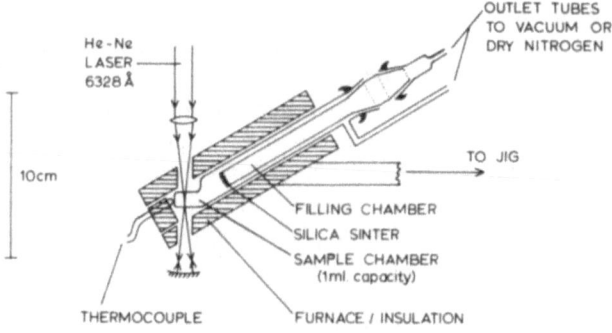

Fig. 4. Molten sample cell and furnace assembly.[11]

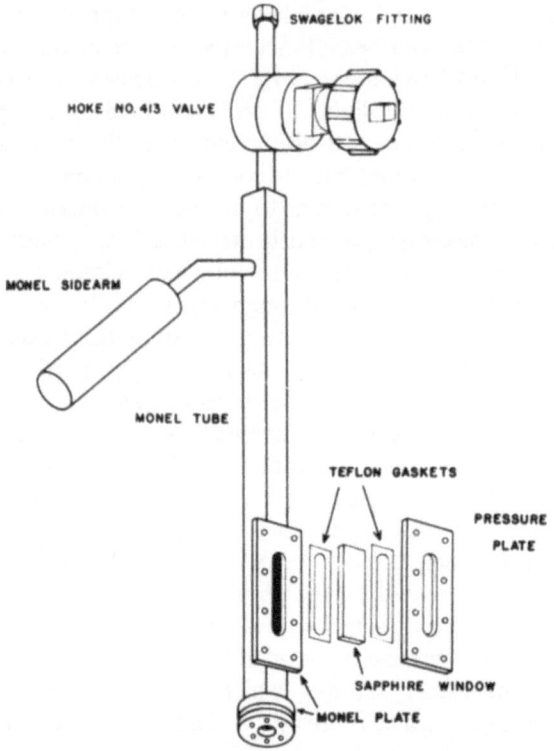

SWAGELOK FITTING

HOKE NO. 413 VALVE

MONEL SIDEARM

MONEL TUBE

TEFLON GASKETS

PRESSURE
PLATE

SAPPHIRE WINDOW

MONEL PLATE

Fig. 5. Raman cell for corrosive liquids.[12]

that their design could be adapted favorably to laser-excitation techniques. This cell, shown in Fig. 5, is made of Monel, with sapphire
windows sealed by Teflon gaskets located at the side and bottom, and
has a volume of about 3 cc. Another promising design has been suggested by Solomons, Clarke, and Bockris,[13] who have used a windowless
Raman cell made entirely from boron nitride, a very inert material
which is easy to fabricate and is relatively inexpensive. Figure 6 shows
the essential features of this cell, which relies on surface tension to retain
the liquid sample in the laser beam, and is similar in concept to a cell
used by Young[14] for absorption spectrophotometry. Spectra of molten
fluorides up to 1030°C have been obtained with this cell. Another
assembly well suited to highly corrosive samples, for which silica
containers are unsuitable, is that of Bues,[9] wherein the sample is
contained in a platinum boat (or other inert material suitable for use at

high temperatures) in a simple crucible furnace. Only slight modifications of this early design, which Bues used with mercury arc-excitation techniques, would be required for laser excitation.

The special characteristics of the Cary 81 laser-Raman spectrometer sample-excitation method, wherein a coaxial optical arrangement is used to produce backscattered (180°) Raman radiation, have formed the basis for the design of a high-temperature cell by Melveger and coworkers.[5] This is adapted from the standard Cary powder-sampling rod, with a solid sample packed in the grooved end to permit heating by a nichrome wire winding, and is shown in Fig. 7. Spectral changes accompanying the *ca.* 140°C phase change from red to yellow solid mercuric iodide have been recorded using this cell. The same authors[15] have also reported a simple U-tube liquid cell, internally heated by nichrome wire, for use with the Cary 81 optical arrangement. This has produced excellent quality spectra from liquid HgI_2 (mp; 257°C), allowing determination of a band as low as 17.5 cm^{-1}. A simple nichrome wire-heated cell has also been reported by Beattie[16]

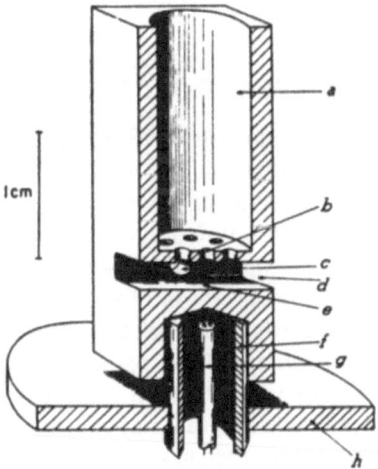

Fig. 6. A section of the windowless Raman cell used for molten cryolite; a) filling chamber; b) retaining holes; c) laser-beam entry hole; d) hole for emerging scattered light; e) sample chamber; f) quartz tube support; g) thermocouple leads; h) boron nitride disk.[13]

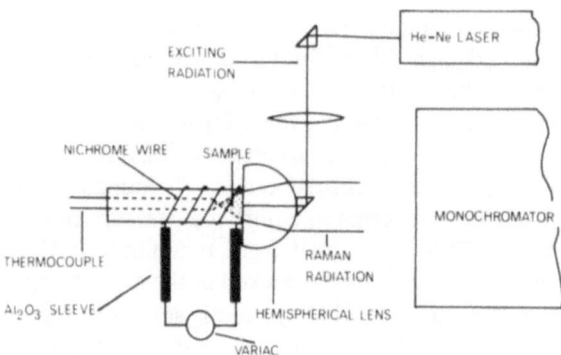

Fig. 7. High-temperature, laser-Raman, solid-sampling technique. Laser excitation is at 180°. Nichrome wire carrying
current serves as heater.[15]

for use specifically with the Cary 81 laser-spectrometer. One unfortunate feature of the Cary 180° scattering arrangement, using a hemispherical lens to collect the Raman radiation, is the difficulty in
obtaining accurate depolarization ratios.

LOW-TEMPERATURE TECHNIQUES

Low temperature sample cells for Raman spectroscopy had been
developed to a greater extent than those for high-temperature work
before lasers were introduced as light sources. There are, accordingly,
many more designs in the literature of cooled cells for use with mercury
arc lamps. Ferraro[3] has reviewed liquid-cell designs by himself and
his coworkers[16] and Griffiths,[17] both of which used vacuum-jacketed
Wood's tubes cooled by flowing precooled, dry nitrogen gas around the
sample tube. A similar cell, but using a solid Pyrex light pipe in place
of the usual double window with vacuum space, has been used by Craig
and Overend[18] and is shown diagrammatically in Fig. 8. This cell, and
its surrounding Dewar, is mounted in a close-fitting, slotted lucite sleeve
supported by the lens tube of the foreoptics in a Cary 81 spectrometer.
The authors have suggested that a similar design might also be used for
high-temperature work. Kittelberger and Hornig[19] have achieved
temperatures as low as −195°C with the cell shown in Fig. 9, which also
uses a precooled nitrogen-gas stream, but which has the extra advantage of being useful for work with HF since it uses a fused quartz inner

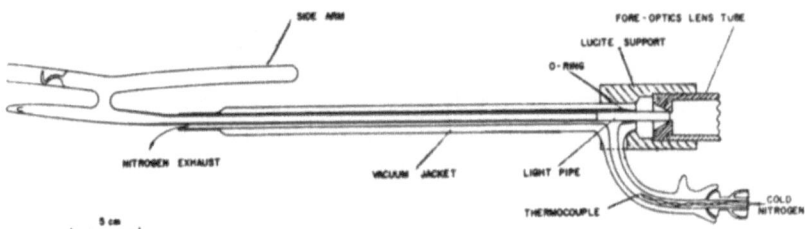

Fig. 8. Outline of Raman cell, vacuum jacket, and support.[18]

tube to contain the sample. An ingenious cell specifically designed for crystalline solids at low temperatures is shown in Fig. 10. This was devised by Blumenfeld and Fast[20] for a comparative study of the low-frequency Raman spectra of solid and liquid formic acid, and was used down to −100°C. The 11-mm-i.d. pyrex tube has a polished window at the lower end and an S-shaped light trap at the other. Clear solid was formed in the tube by means of refrigerating the whole assembly and then slowly (<0.5 mm/h) raising the external coil heater, which kept the solid–liquid interface inclined to the plane of the window, ensuring that the crystal was nucleated at a single point on the rim and grew in one direction across the window, thus eliminating any faults. The tube was then transferred to a cryostat consisting of a double-windowed Pyrex Dewar, cooled on the inside by a flow of nitrogen.

The nitrogen-flow cooling technique used in all the above examples has been criticized as expensive and inconvenient by Bryant,[21] who has

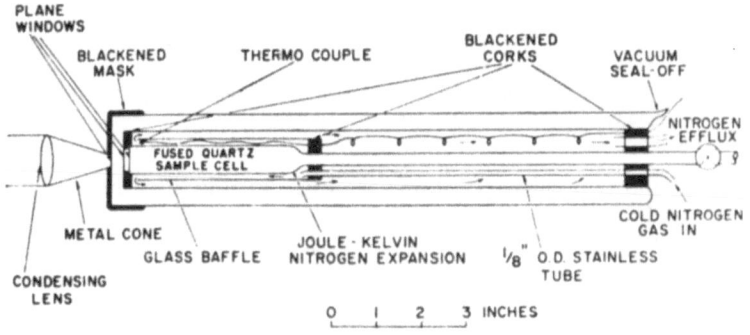

Fig. 9. Low-temperature Raman apparatus.[19]

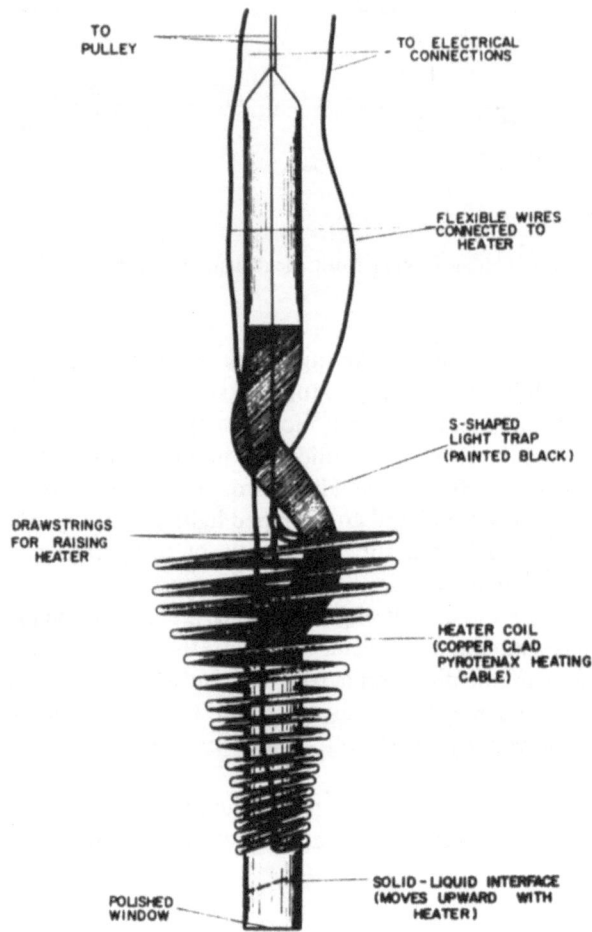

TO
PULLEY

TO ELECTRICAL
CONNECTIONS

FLEXIBLE WIRES
CONNECTED TO
HEATER

S-SHAPED
LIGHT TRAP
(PAINTED BLACK)

DRAWSTRINGS
FOR RAISING
HEATER

HEATER COIL
(COPPER CLAD
PYROTENAX HEATING
CABLE)

POLISHED
WINDOW

SOLID-LIQUID INTERFACE
(MOVES UPWARD WITH
HEATER)

Fig. 10. Schematic diagram of apparatus used for growing formic
acid crystals in the Raman tubes.[20]

devised the two cryostats shown in Figs. 11 and 12, wherein liquid
nitrogen is used directly as a coolant, giving sample temperatures in the
region of $-190°C$. System A is for use with large, flat crystals, and B is
for smaller but bulkier crystalline samples. The samples are mounted on
the metal block at the end of each tube, and are run as solids. Still
lower temperatures have been achieved by use of liquid helium as a
coolant in the Raman cryostat used by Savoi and Anderson.[22] Their

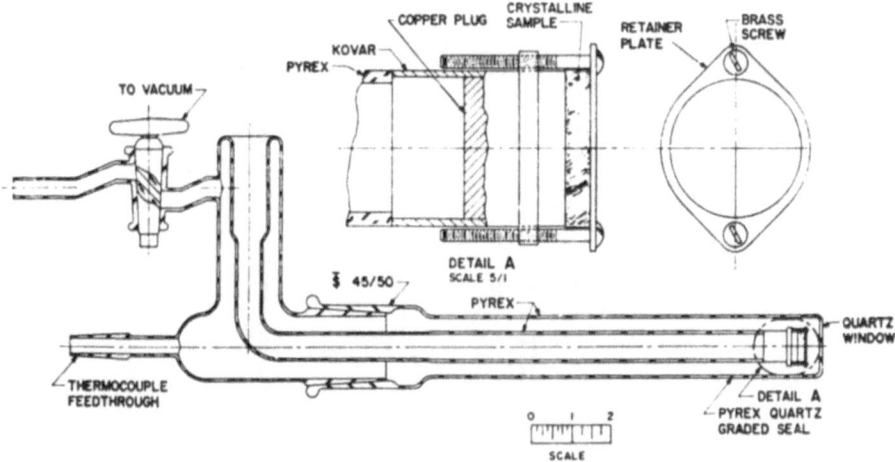

Fig. 11. Cryostat for low-temperature Raman spectroscopic measurements. Design A.[21]

assembly, which is shown diagrammatically in Fig. 13, enabled sample temperatures in the region of 13°K to be reached, allowing samples which are normally gases at room temperature to be run as solids. An entering sample is first cooled to a liquid in this system, and then further cooling produces a "lollipop"-like crystal growth around the polished copper rod. The earlier liquid-helium cryostat designed for Raman spectroscopy with Toronto-arc excitation by Bhatnager, Allin, and Welsh[23] is evidently more efficient, however, since it has enabled samples to be run[23-25] as low as 2.1°K, the λ point of liquid helium.

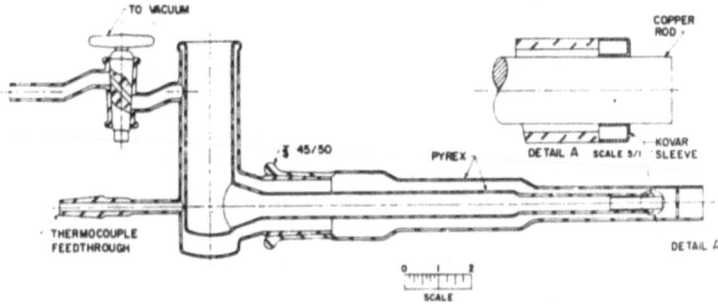

Fig. 12. Cryostat for low-temperature Raman spectroscopic measurements. Design B.[21]

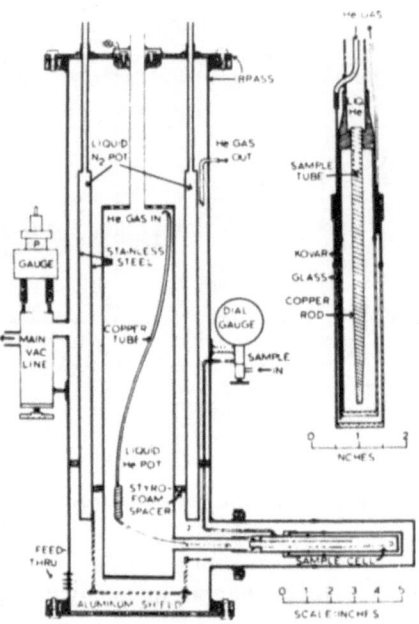

Fig. 13. Liquid-helium Raman cryostat.[22]

Figure 14 shows the essential features of this Raman cryostat, which is
built around coaxial, cylindrical Dewars, fused together to form a single
four-walled vacuum flask. The outer compartment (A) holds liquid
air, the inner one (B) liquid helium. Plane Pyrex windows allow exit of
the scattered light from the Pyrex sample tube (F) after passage through
the plane window (J) and deflection by the prism (L), which is connected
by the Pyrex tube (K). A brass spacer (M) holds the sample tube in
position, and the sample area is surrounded by an outer filter jacket (C).
A heater (E) can be used to raise the temperature by boiling of helium.
The whole assembly is located within the coils of the mercury lamp (D).
A high-speed pumping unit is necessary to lower the helium temperature
to the λ point and to prevent bubbling, which would cause intensity
fluctuations in the light incident on the sample. The various forms of
hydrogen have been run as both solids and liquids in this cell.

It is clear that several of the preceding cell designs could be simpli-
fied for use with laser excitation. Figure 15 shows a system used by
Carlson[26] with the Cary 81 laser instrument. The cell is made of
Pyrex glass, with the sample being held at the end of a brass rod and

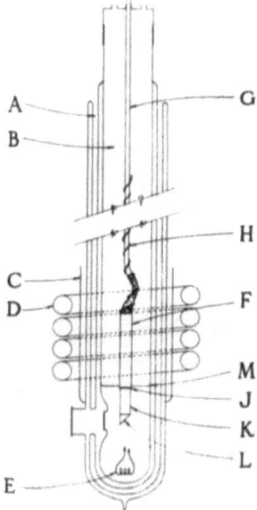

Fig. 14. Vertical cross section of the arrangement for Raman spectroscopy at liquid-helium termeperatures.[23]

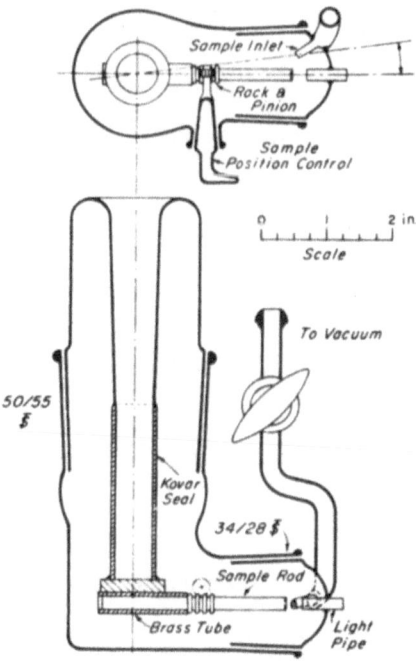

Fig. 15. Low-temperature Raman cell.[26]

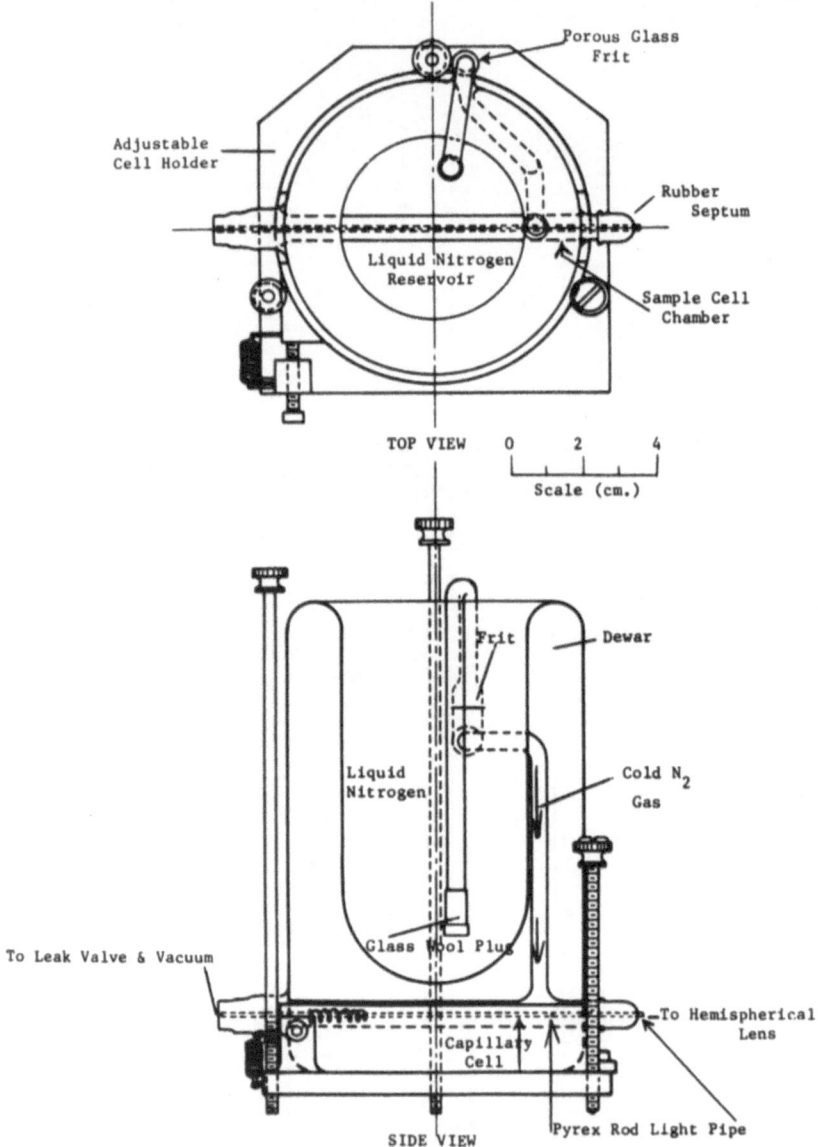

Fig. 16. Low-temperature cell for the Cary 81 laser-Raman instrument.[27]

coupled to the Cary hemispherical lens by a solid light pipe. Using liquid nitrogen as the coolant, a sample temperature of −135°C was achieved with this cell, indicating rather poor thermal contacts. Carlson and Fateley[27] have constructed a more successful cell, enabling spectra to be obtained down to −175°C. This cell is also peculiarly suited to the Cary 81 optical system employing the coaxial viewing technique, and is shown in Fig. 16. The sample (liquid or solid) is sealed into a 1-mm-i.d. capillary tube which has a 2.5-cm-long Pyrexrod light pipe fused to the window end for coupling to the hemispherical lens. This sample tube is supported in the Dewar chamber by a rubber septum and a coiled wire spring, and is cooled by cold N_2 gas pulled through from the Dewar reservoir. Another cell made from Pyrex capillary tubing, with a volume of 0.1 cm^3, has been described by Carpenter and coworkers.[28] Their cell was cooled to −70°C in a special Dewar cell holder by an acetone–dry ice mixture, and used with a focussed, argon-ion laser beam, the scattered light being collected through the side of the tube.

One of the most generally useful Raman cryostats designed for laser work with single crystals appears to be that of Gee and O'Shea.[29] Their system, shown in Fig. 17, is made entirely of Pyrex glass which is all silvered, except for a small slot at the sides and bottom of the tip. A slotted aluminium tube holds the crystal in place at the bottom of the central column, which is flushed by nitrogen gas evaporating in the Dewar. The cold gas stream both cools the crystal and prevents condensation of moisture within the sample area. A cw laser beam is directed vertically into the tip from the bottom, and scattered light is collected at the side. The dimensions of this cryostat are not critical, but the original assembly is 9 cm in diameter and 44 cm in length, with a capacity of 1 liter, allowing 18 h operation with a single filling. The crystal temperature in this system was found to be only 1° above that of the liquid nitrogen, and it would seem relatively easy to add a further nitrogen jacket to allow liquid-helium temperature to be reached. A further development of this for liquid-helium work has been reported by Gee and Robinson.[29a]

It can be seen that a variety of designs can be utilized for low-temperature Raman spectroscopy, and others probably will be presented in the literature before this book appears in print, particularly where laser excitation is used. A single-crystal laser-Raman cryostat for use at liquid-helium temperatures already was mentioned by Koningstein.[30]

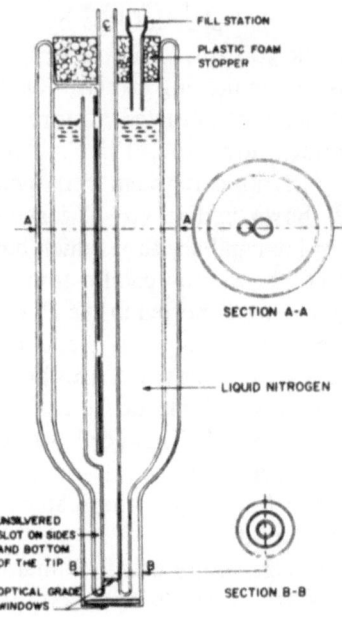

Fig. 17. Section of a Dewar for use in crystal spectroscopy.[29]

MOLTEN SALTS AND OTHER HIGH-TEMPERATURE SYSTEMS

Most of the activity in the area of high-temperature Raman spectroscopy has been concerned with molten salts, and this topic has already been very adequately reviewed by Janz and Wait.[7] However, significant further progress has been made during the past two years, particularly with respect to systems containing polyatomic ions. Here, as in most vibrational spectroscopic work, considerable benefits accrue from parallel studies of infrared and Raman spectra under similar conditions. Such studies of the zinc nitrate–water system, for example, have been made,[31] where high-temperature techniques enabled the system to be investigated in the liquid state from the simple aqueous-solution region through to the molten salt hydrate of composition $Zn(NO_3)_2 \cdot 1.42H_2O$. Variable temperature and variable water-content studies of the spectra of this system showed strong evidence for endothermic displacement equilibria of the type

$$Zn(H_2O)_6^{+2} \underset{H_2O}{\overset{NO_3^-}{\rightleftarrows}} Zn(NO_3)(H_2O)_5^+ \underset{H_2O}{\overset{NO_3}{\rightleftarrows}}$$

$$Zn(NO_3)_2(H_2O)_4 \underset{+H_2O}{\overset{-H_2O}{\rightleftarrows}} Zn(NO_3)_2(H_2O)_{<4}$$

with the species on the right becoming progressively more important with decreasing water content. The Raman polarization data were used to establish that the inner-sphere-coordinated nitrate ions in this system are attached to the zinc ion through a single oxygen atom, thus lowering the NO_3^- ion symmetry from D_{3h} (the "free" ion symmetry) to C_{2v}.[32,33] Although thermal decomposition resulted from attempts to further dehydrate the molten $Zn(NO_3)_2 \cdot 1.42H_2O$, preventing spectra from the anhydrous molten salt being obtained, the spectral trends with decreasing water content were interpreted as showing that this would probably be essentially a fully associated liquid. A dramatic loss of symmetry of the nitrate environment on melting the hexahydrate was indicated by the spectra, suggesting breakup of the symmetrical hexaquo-zinc ion complexes[34] by inner-sphere nitrate coordination.

Further studies of the Raman spectra of doubly-charged metal nitrates in the anhydrous state as solutions in or mixtures with alkali metal nitrates have been made[35] and confirm the earlier infrared absorption results for such systems.[36] It appears that strong perturbations of the nitrate ion vibrational spectrum are caused by association with most multiply-charged metal ions in anhydrous molten salt systems, and that the Raman and infrared spectra of this and other such polyatomic ions can be used as probes in determining the structural features of molten salts. Raman spectra of mixed nitrate glasses[37] provide a further illustration of this effect, and offer the possibility of a wide temperature range over which to study the structure of such ionic liquid systems.

A Raman study by James and Leong[38] of sodium nitrate as a crystalline solid and as a melt at various temperatures has shown evidence for the presence of crystallites in the liquid just above the melting point, with a packing resembling that of the crystal, although it also has been suggested that the quasicrystalline order in the melt is different from that of the crystal.[37] High-resolution infrared spectra of alkali metal nitrate melts have similarly produced evidence for a latticelike structure in the liquid,[39] and frequency differences between the Raman and infrared bands of molten $ZnSO_4$–K_2SO_4 mixtures have been discussed[40] in terms of correlation field-splitting effects

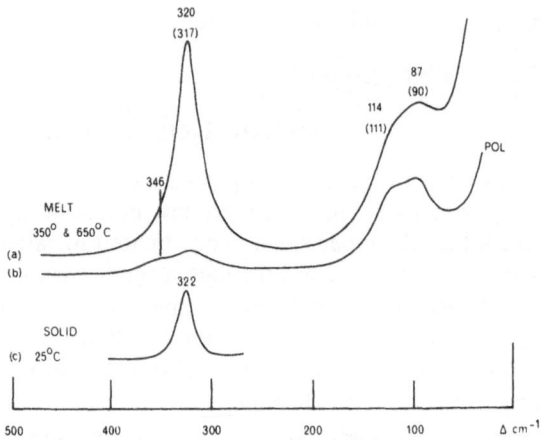

Fig. 18. Raman spectra of $InCl_3 \cdot KCl$. Frequency shifts of band maxima are in cm^{-1}; figures in parentheses refer to 650°C. POL indicates trace obtained with the polarizer in the scattered beam and transmitting only light with its electric vector perpendicular to that of exciting light. The intensity scales of different spectra are not comparable.[42]

arising from coupling of vibrational modes between neighboring sulfate ions in a limited, quasicrystalline liquid structure. However, for melts containing multiply charged metal ions, at least, the experimental evidence appears to favor the existence of distinct metal–anion complexes being responsible for the gross perturbations of the polyatomic anion spectra which are observed, with latticelike liquid interactions giving rise to secondary effects.

Raman spectra from metal halide melts appear uniformly to have been most satisfactorily interpreted in terms of complex-ion formation. A recent example is provided by Clarke and Solomons,[41] who obtained spectra from molten and crystalline stannous chloride and molten mixtures of stannous chloride and potassium chloride at temperatures from 250 to 600°C. These produced evidence for the pyramidal species $SnCl_3^-$ in the mixtures with excess KCl, though with low KCl content or the pure $SnCl_2$ melt chain polymers of the type $(SnCl_2)_n$ were indicated, with 3-coordination of Sn(II) being preserved. In molten cryolite, Na_3AlF_6, at 1030°C, Solomons, Clarke, and Bockris[13] also found evidence for complex ions. Their Raman spectra, obtained using the captive liquid cell described earlier, gave bands assignable to both tetra- and hexafluoroaluminate ions, with the relative band intensities indicating a degree of dissociation of the AlF_6^{-3} species into AlF_4^- and $2F^-$ ions somewhere between 0.8 and 0.6.

The high quality which can be obtained in laser-Raman spectra of high-temperature liquids is illustrated in Fig. 18, which shows[42] a four-line spectrum characteristic of the tetrahedral species $InCl_4^-$, where the signal-to-noise ratio observed for the main peak ($320\, cm^{-1}$) was in excess of 150. It is remarkable that the band frequencies and relative intensities in this spectrum are closely similar to those obtained from $InCl_4^-$ dissolved in ether at normal room temperature.[43] A similar observation has been made for the $SnCl_3^-$ spectrum.[41] Raman spectra also have been used to establish the structure of $InCl_2$[44-46] in the molten state as $In^+(In^{III}Cl_4)^-$, and of molten In_2Cl_3,[46] as $In_3^+(In^{III}Cl_6)^{-3}$, though the spectra show that considerable dissociation of the $InCl_6^{-3}$ species occurs in the melt, and some chloride-induced disproportionation of In^+ to In^0 and In^{+3} ensues.[46]

Further progress in determining the nature of molecular melts has been made through the Raman studies of Clarke and Solomons[47] and

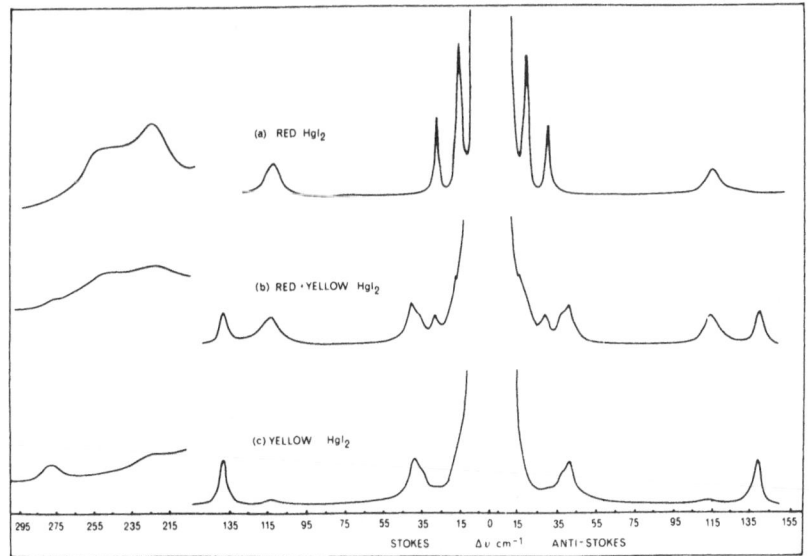

Fig. 19. a) Raman spectrum of room-temperature red HgI_2. b) Raman spectrum of the red and yellow HgI_2 mixture at the point of incomplete phase conversion. c) Raman spectrum of yellow HgI_2 just above the transition point. Residual trace of red HgI_2 is indicated.[15]

Melveger and coworkers.[15] Earlier work on molten mercuric chloride, bromide, and chlorobromide by Janz and James[48] has been extended with the aid of laser excitation to the deeply colored iodide. The linear triatomic structure of HgI_2 is evidently preserved in the molten state, and comparison of frequencies for the v_1 symmetric stretching modes of this and other HgX_2 compounds when measured in the vapor[49] and liquid states indicates a relatively high covalent character for the Hg–I bonds.[47] The solid-state phase transition from the red to the yellow form of HgI_2 (at 126°C) also has been examined through changes in the Raman spectrum,[15] and the results interpreted in terms of the nonmolecular structure of the red form and the molecular structure of the yellow form. In Figs. 19 and 20 the Raman spectra from the various forms of HgI_2 are reproduced,[15] showing, in Fig. 19, that strikingly different spectra are given by the two solid forms, and, in Fig. 20, that the low-frequency doublet given by the yellow form gives way to a singlet on melting. It is believed that the doublet (at 37 and 41 cm^{-1}) is due to the normally Raman-inactive E_u modes of molecular HgI_2,

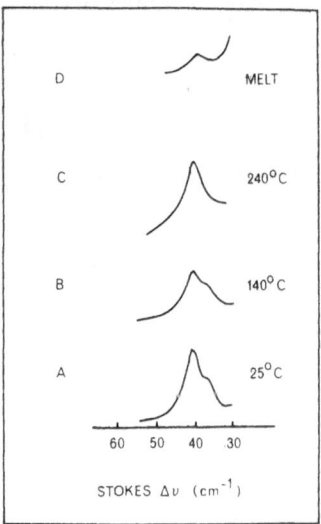

Fig. 20. A) Spectrum of the low-frequency doublet of metastable yellow HgI_2 at room temperature. B–D) Progressive increase in temperature removes the doublet structure but the single band is still present in melt.[15]

with the degeneracy lifted and selection rules relaxed by crystal-lattice perturbations. The persistance of the 41-cm^{-1} line in the melt indicates a breakdown in selection rules due to less well understood environmental effects.

RESULTS FROM LOW-TEMPERATURE STUDIES

Until recently, obtaining a Raman spectrum of a gaseous sample has been an elaborate and time-consuming business, and chemists have commonly resorted to cooling their samples to form liquids for Raman work. Now that it is possible to get good quality Raman spectra from gases merely by blowing them at atmospheric pressure through a laser beam,[50] it is likely that less attention will be given to low-temperature studies by those whose interest is mainly in determining the molecular structure of the samples. However, low-temperature techniques will continue to be of interest in a broad range of topics where the liquid state of low-boiling substances is needed and, of course, for crystal studies.

Results obtainable from the combination of low-temperature techniques with a modern, high-resolution, laser-Raman spectrophotometer are well illustrated by the chlorine-isotope splitting patterns for the v_1 (A_1) fundamentals of solid CCl_4, $SiCl_4$, $GeCl_4$, and $SnCl_4$ shown in Fig. 21.[51] These were obtained from sublimed films at about 80°K, and show relative line intensities in fair agreement with the theoretical ratios, viz., XCl_4^{35}, 81; $XCl_3^{35}Cl_1^{37}$, 108; $XCl_2^{35}Cl_2^{37}$, 54; $XCl_1^{35}Cl_3^{37}$, 12; and XCl_4^{37}, 1. Another example of a low temperature Raman spectrum is shown in Fig. 22,[21] in which are reproduced the spectra of a single crystal of silver azide in the symmetric stretching frequency region (1200–1380 cm^{-1}) at 313 and 80°K. These demonstrate the usefulness of low-temperature spectra in making assignments of spectral bands, and have enabled Bryant[21] to show that for AgN_3 there is a Fermi resonance interaction between the modes v_1 (1347 cm^{-1}) and $2v_2$ (1257 cm^{-1}), and that the further complication of the region is due to correlation field splitting of the two resonant modes. Factor-group selection rules predict that two components for v_1 $(a_g + b_{1g})$ and four components for $2v_2$ $(4a_{1g})$ should be Raman active, and the good resolution achieved in the 80°K spectrum enabled identification of the individual components of v_1 at 1327 and 1347 cm^{-1} and of $2v_2$ at 1246, 1257, 1264, and 1289 cm^{-1} to be made.

Variable-temperature techniques for the study of vibrational

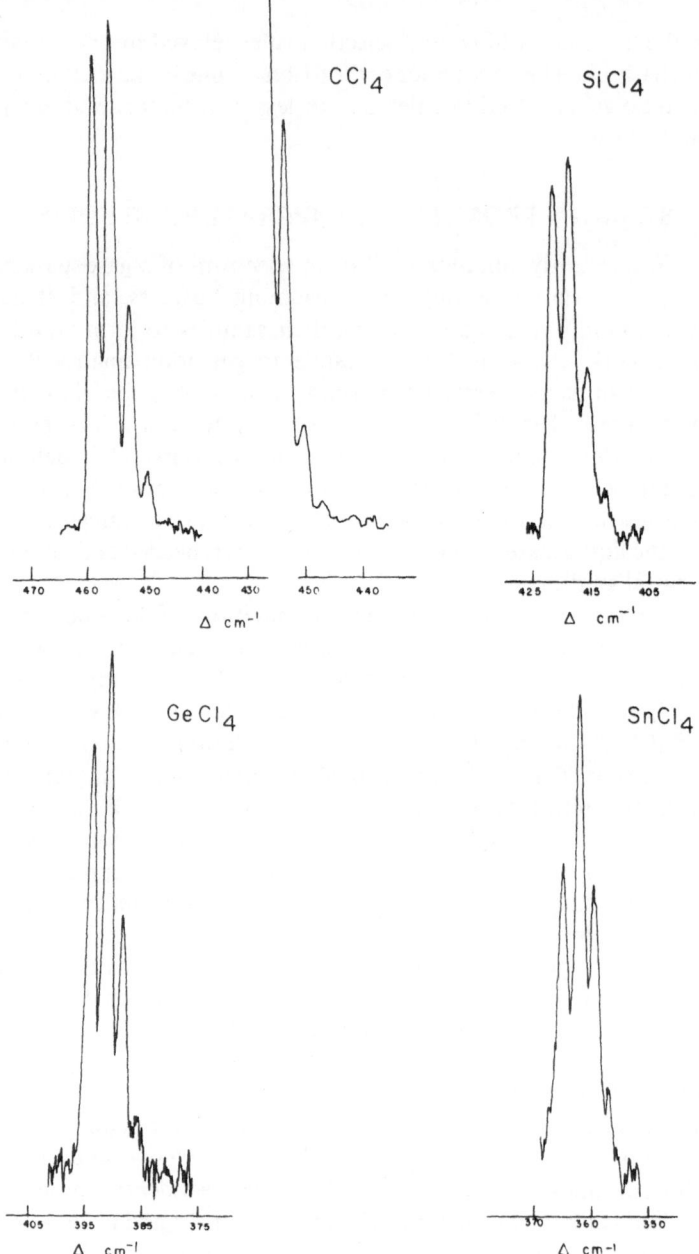

Fig. 21. Chlorine isotope splitting patterns for v_1 (A) fundamentals of CCl_4, $SiCl_4$, $GeCl_4$, and $SnCl_4$.[51]

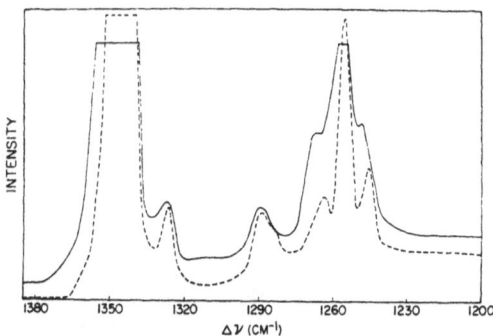

Fig. 22. Raman spectra of the symmetric stretching frequency region of a single crystal of silver azide.[21]
——— 313°K, – – – – 80°K.

spectra enable phase-change phenomena to be investigated through the effects on the frequencies and intensities of characteristic modes. Perzl and Moser[52] have studied the changes in absolute intensities of several Raman lines from *t*-butanol which accompany the transitions from gas → liquid → solid, and their results are important for the information which they provide on intermolecular interactions in the condensed phases, and also in the general context of interpretation of Raman intensities from condensed phases.[53] They found amplification factors of the absolute intensities for the gas → liquid phase transition were higher for polarized lines (factor 3.3–3.5) than for depolarized lines (factor 1.3–1.8), but that a uniform amplification (factor 2.0–2.2) accompanied the liquid → crystalline solid transition. In some earlier work by Bhatnagar, Allin, and Welsh,[23] it was found that Raman line *frequencies* were lowered by the phase transition of gas → liquid (by 7–9 cm⁻¹) and gas → solid (8–11 cm⁻¹) in hydrogen. Raman work on liquid and solid H_2, D_2, and HD at high resolution,[23] on the frequency and band shape dependence on the ortho/para ratio in solid hydrogen,[24] and on an intensity anomaly in the spectra of solid and liquid hydrogen[25] also has been reported by the Toronto group.

Low-temperature spectra of the hydrogen halides HCl, DCl, HBr, and DBr have been reported by Savoi and Anderson.[54] As crystalline solids in the region 77–10°K the Raman spectra showed four components in the stretching region, indicating nonplanarity of the zigzag chains which form these crystals. Subsequent work on crystalline HF and DF at 80°K by Kittelberger and Hornig[19] showed an additional

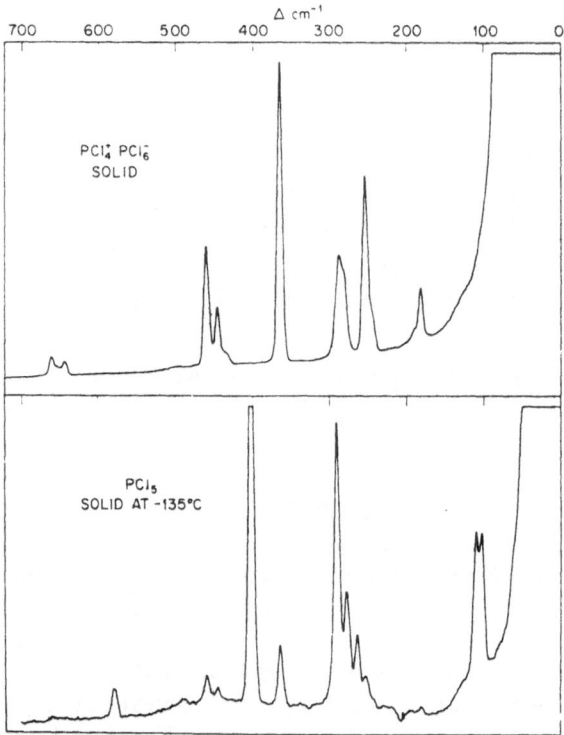

Fig. 23. Raman spectra of solid phosphorus pentachloride demonstrating changes on going from covalent solid to ionic solid.[26]

low-frequency, Raman-active lattice transition, and established the coincidence of Raman and infrared frequencies from these crystals. This coincidence showed the presence of a disordered chain orientation in a D_{2h}^{17}-*Bmmb* crystal structure, rather than the ordered D_{2h}^{16}-*Pmnb* structure (where the mutual exclusion principle is operative).

It is well known that in the solid state phosphorus pentachloride normally has the structure $PCl_4^+ PCl_6^-$. However, rapid condensation of PCl_5 vapor on a surface cooled to liquid-nitrogen temperature produces a covalent solid. This has been demonstrated by Raman spectra obtained by Carlson,[26] who showed that raising the temperature of the covalent solid much above $-135°C$ resulted in spectral changes indicative of the disproportionation reaction. Carlson's spectra are shown in Fig. 23. A wide range of phosphorus halides and mixed halides

have been investigated by low-temperature Raman techniques by
Griffiths, Carter, and Holmes,[55] who studied PF_5, PF_3Cl_2, PF_2Cl_3,
and $PFCl_4$ as liquids. Figure 24 shows both the infrared and Raman
spectra of $PFCl_4$, which has been found to have a distorted trigonal-
bipyramidal structure with C_{3v} symmetry.

Gasner and Claasen[56] have reported Raman spectra from solid
and liquid XeO_2F_2, showing that this noble-gas compound has a
pseudobipyramidal structure of C_{2v} symmetry, with the two fluorine
atoms axial to Xe and two oxygen atoms and a lone pair equatorial.

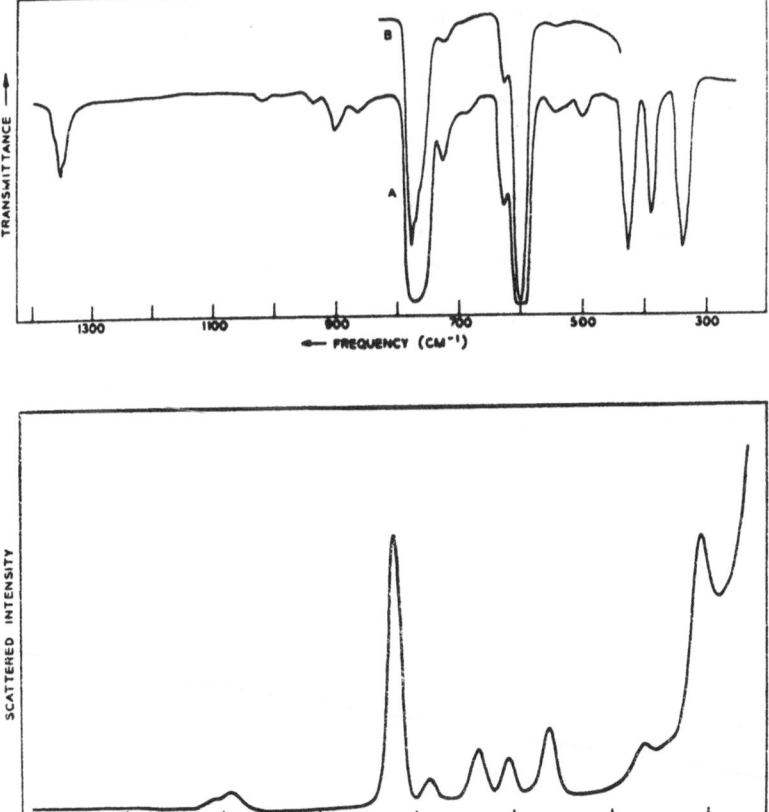

Fig. 24. Infrared and Raman spectra of $PFCl_4$. Infrared (top): A) $p \approx 10$ cm, $1 = 10$ cm;
B) $p \approx 2$ cm, $1 = 10$ cm; temperature, 25°C. Raman (bottom): slit, 10 cm^{-1}; ampl.,
1000; temperature, −40°C; single beam.[55]

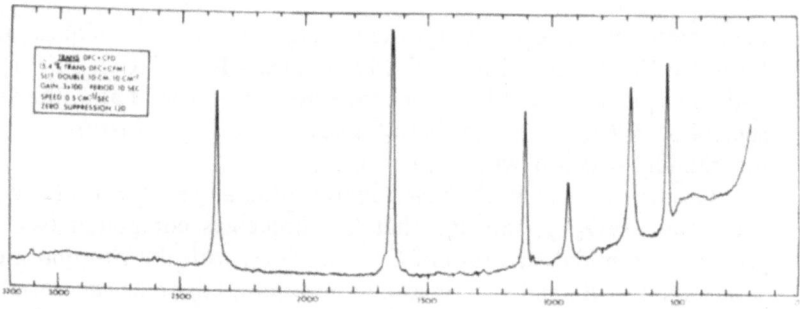

Fig. 25. Raman spectrum of *trans*-DFC–CFD.[18]

A monomeric structure was indicated for both solid and liquid. Another normally gaseous fluorocompound whose Raman spectrum has been obtained from the liquid is *trans*-1,2-difluoroethylene-d_2 (*bp*, $-53°C$). This is shown in Fig. 25.[18] The methyldiboranes provide a further example of the value of low-temperature techniques, since these compounds tend to disproportionate very rapidly at room temperature. Their Raman spectra have been obtained at $-70°C$ by Carpenter and coworkers.[57]

Finally, it is worth mentioning that studies of the optical properties of crystals gain much from investigations at low temperatures. As an example, the work of Johnston and Kaminow[58] on the temperature dependence of Raman and Rayleigh scattering in $LiNbO_3$ and $LiTaO_3$ has established that the ferroelectric transitions in these materials are second order, and are associated with an optic phonon mode whose frequency becomes small as the Curie temperature is approached from below. Both of these materials exhibit large, nonlinear optic effects and are ferroelectric at room temperature. By cooling a crystal of MnF_2 below its Neel–Curie temperature (70°K), at which it becomes antiferromagnetic, Fleury and coworkers[59] have been able to demonstrate Raman scattering by spin waves or magnons. FeF_2 similarly becomes antiferromagnetic at 70°K, and Raman scattering from both one and two magnon states has been observed[59,60] below this temperature. Detailed evidence for the identification of the one-magnon process in FeF_2 was obtained by investigation of the temperature dependence of the scattered light over the range 70–10°K. Electronic Raman scattering by the neutral acceptors Zn and Mg in GaP (these consisting of a hole trapped on a Group-II impurity) also has been reported[61] from samples held at 20°K. Even molecular crystals such as C_6H_6

and C_6D_6 have yielded interesting new information on intermolecular coupling of vibrations through Raman studies at 77 and 2°K, where splitting of each of the degenerate fundamental modes was observed.[29a]

ASSESSMENT OF RESULTS

It is seen that most of the results from Raman spectroscopy have been obtained from condensed phases—either liquids or solids. Liquid spectra have usually been interpreted in terms of the intramolecular vibrational modes of the polyatomic species present, with but little attention being paid to the possible effects of intermolecular perturbations. Molecular solids have commonly been treated in a similar way, though some attempts have been made to assign lattice modes. Spectra from ionic crystals, on the other hand, have uniformly and necessarily been interpreted on the basis of the phonon model. The class of liquids derived from fusion of ionic salts falls into a special category, since much evidence has been accumulated to show the presence of distinct molecular complexes in such systems and yet it seems probable that some partially disordered form of lattice structure might be retained, at least at temperatures not far above the melting point. The selection of examples to illustrate the problems here is difficult, since each system introduces its own bias. Wilmshurst[62] has made an attempt to establish vibrational spectroscopic criteria for the existence of kinetic complex species in molten salts, but concluded that, since the lifetime of a whole system configuration in ionic liquids, or even in aqueous solutions, is likely to be of the order of 10^{-10}–10^{-11} sec, only ambiguous information can in general be obtained from such systems. The implications here seriously affect many of the conclusions made on the basis of results of Raman and infrared, and indeed also of UV/visible absorption studies, and are worth investigating further.

The system HgI_2 discussed earlier is a useful one for this purpose, since as a solid it exists in both an ionic and a molecular form, and it has also been studied in the molten state by Raman spectroscopy.[15,47] Thus, it can serve to illustrate the methods used for treating vibrational modes appropriate to each of these circumstances, and experimental results are available for comparison with the theoretical predictions.

X-ray crystallographic work[63] has shown that crystals of yellow HgI_2 are orthorhombic, with space symmetry $Cmc2_1(C_{2v}^{12})$. The unit cell has four molecules arranged as shown in Fig. 26. This is the same arrangement as in the orthorhombic $HgBr_2$ crystal,[64] and is based on

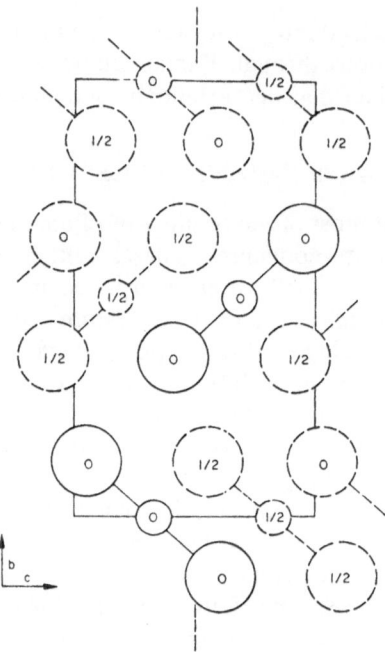

Fig. 26. Crystal structure of mercuric
bromide. A projection along a_0 axis. The
large circles are bromine atoms. The
figures in circles indicate the coordinates
in the a direction.[64]

a severely distorted hexagonal, close-packed structure. Within the
HgI_6 octahedra two Hg–I distances are 2.617 Å, which is approximately
equal to the sum of the Hg and I covalent radii, while the other four
are 3.508 Å, which is 0.1 Å shorter than the sum of the usually accepted
van der Waals radii.[63] Thus, the structure may be considered essentially
molecular, though the I–Hg–I angle is 178.3 ± 0.3°, showing a small
but significant deviation from linearity. Two molecules of the unit cell
are in the bc plane (see Fig. 26) and the other two are a half-unit above
or below this plane. Only two molecules need be considered for
determination of the optically active modes,[65] since they are related to
the other two by a simple translation. The 18 degrees of freedom which
then arise can be classified by the usual group theoretical methods as
$6A_1 + 4A_2 + 2B_1 + 6B_2$.[64] Since the intermolecular forces are likely
to be weak compared with the intramolecular forces, each vibrational

Table 1. Character Table and Classification of Vibrational Modes of the HgI$_2$ (Yellow) Crystal

C_{2r}^{12}	E	$C_2(b)$	$\sigma_g(ab)$	$\sigma(bc)$	n	A	T	L	n_i
A_1	1	1	1	1	6	1	1	1	3
A_2	1	1	-1	-1	4	1	1	1	1
B_1	1	-1	1	-1	2	0	0	1	1
B_2	1	-1	-1	1	6	1	1	1	3

n is the total number of modes; A, T, L, and n_i are, respectively, the number of acoustic, translational librational, and internal modes.

Table 2. Correlation Table for the Factor Group, Site Group, and Point Group of the HgI$_2$ (Yellow) Crystal

Factor group C_{2r}	Site group Cs	Point group $D_{2\infty}$
$A_1(v_1, v_2, v_3)$:R, IR	A'	$\Sigma^+(v_1)$:R
$A_2(v_2')$:R		$\Sigma^-(v_3)$:IR
$B_1(v_2')$:R, IR	A''	
$B_2(v_1, v_2, v_3)$:R, IR	R, IR	$\Pi(v_2)$:IR

R indicates Raman active; IR, infrared active.

mode can further be classified as an acoustic, internal, or external (translational or librational) mode.[66,67] This classification is shown in Table 1. The correlation between the factor group, the site group, and the point group of the molecule is given in Table 2, together with the activities of the various species in the Raman and infrared spectra. It is clear that in the limit of no intermolecular interaction the eight internal modes will reduce to the three normal modes, v_1, v_2, and v_3, of the free HgI$_2$ molecule. Even with moderate interaction, the frequencies v_1 (A_1) and v_3 (A_1) should be very similar to v_1 (B_2) and v_3 (B_2), respectively, since the corresponding modes differ only in phase. It can be anticipated, therefore, that these modes will appear as doublets (perhaps unresolved) at frequencies close to those of the free (gaseous) molecule. However, the doubly degenerate bending mode of the free molecule, v_2 (π), should be more strongly split in the crystal spectrum, since the large separation between the HgI$_2$ molecular layers will distinguish the in-plane mode v_2 from the out-of-plane mode v_2'. In accord with these predictions, the Raman spectrum of yellow HgI$_2$ showed[15] v_1 as a single band at 138 cm^{-1} (cf. 155 cm^{-1} from the gas), and v_2 and v_2' as a weak doublet at 41 and 37 cm^{-1} (cf. 33 cm^{-1}). v_3 remained inactive (very weak) in the Raman effect, but was observed as a fairly strong infrared absorption at 200 cm^{-1} (cf. 237 cm^{-1}).

The red HgI_2 crystal is tetragonal, belonging to the space group $P4_2/(D_{4h}^{15})nmc$, and having two molecules per unit cell.[63] However, this is a predominantly ionic lattice, based on a distorted cubic, close-packed arrangement, with layers of HgI_4 tetrahedra linked at some of their vertices, the Hg–I distances within the tetrahedra being 2.783 Å. The unit cell is shown in Fig. 27; the I atoms are located on sites having $4d$ or C_{2v} symmetry and the Hg atoms on sites of $2a$ or D_{2d} symmetry. The factor-group analysis of this structure produces the following distribution of optical modes:[15] $A_{1g} + B_{1g} + 2E_g + 2A_{2u} + B_{2u} + 3E_u$, with the presence of a center of symmetry resulting in operation of the mutual exclusion role for infrared and Raman activities, and the B_{2u} mode being inactive. Four modes should therefore be active in the Raman spectrum ($A_{1g} + B_{1g} + 2E_g$), and five ($2A_{2u} + 3E_u$) in infrared absorption. The observed Raman spectrum[15] contains a strong line at 114 cm^{-1} which can be assigned to the A_{1g} mode; others

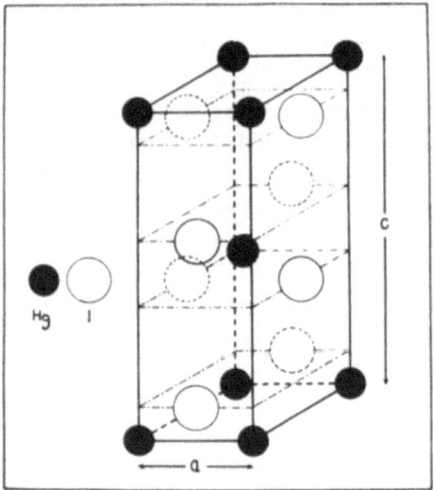

Fig. 27. Tetragonal unit cell of red HgI_2 in the space group $P4_2/nmc$ (D_{4h}^{15}) containing two molecules of HgI_2. Hg atoms are located at sites $2a$ having coordinates $(\frac{3}{4},\frac{1}{4},\frac{3}{4})$; $(\frac{1}{4},\frac{3}{4},\frac{1}{4})$. I atoms are at sites $4d$ having coordinates $(\frac{1}{4},\frac{1}{4},z)$; $(\frac{3}{4},\frac{3}{4},z)$; $(\frac{1}{4},\frac{1}{4},\frac{1}{2}+z)$; $(\frac{3}{4},\frac{3}{4},\frac{1}{2}-z)$, where $z = 0.39$ and the origin is at I.[15]

at 29 and 17.5 cm^{-1} and several broad and weak bands have not been assigned. The A_{1g} mode can perhaps be compared with the v_1 mode of the yellow form of HgI_2, but it is seen that its frequency is very different (114 cm^{-1} compared with 138 cm^{-1}) and still further from the HgI_2 gaseous molecule v_1 frequency (155 cm^{-1}). Clearly, the vibrational spectrum provides an excellent way of distinguishing the two different crystalline forms of HgI_2.

If the structure of molten HgI_2 is molecular it must be expected to give rise to a vibrational spectrum similar to that of yellow HgI_2, or even to that of the gaseous molecule if the liquid structure does not have a latticelike character. On the other hand, an ionic melt should give a very different spectrum, resembling that of red HgI_2 if there is a highly ordered lattice structure in the liquid, but degrading further to broad and very low frequency bands if the structure is not latticelike. The observed Raman spectrum, as stated earlier, shows that the liquid is molecular, with restoration of the degeneracy of the E_u modes indicating break up of the layer–lattice structure of yellow HgI_2, but retention of weak Raman activity of these modes indicating a less specific perturbation arising from intermolecular forces. This type of detailed comparison of spectra from crystals of known structure with the corresponding liquid spectra can contribute much toward an understanding of the nature of intermolecular (interionic) interactions and the structure of liquids, and it is hoped that more variable temperature spectroscopy will be inspired by this review.

REFERENCES

1. R. W. Wood, *Phys. Rev.* **36**:.1421 (1930).
2. H. L. Welsh, M. F. Crawford, T. R. Thomas, and G. R. Love, *Can. J. Phys.* **30**: 577 (1952).
3. J. R. Ferraro, in: H. A. Szymanski, *Raman Spectroscopy*, Plenum Press, New York (1967), Vol. 1, Chapt. 2.
4. G. E. Walrafen, *J. Chem. Phys.* **43**: 479 (1965).
5. G. E. Walrafen, D. E. Irish, and T. F. Young, *J. Chem. Phys.* **37**: 662 (1962).
6. G. J. Janz, Y. Mikawa, and D. W. James, *Appl. Spectry.* **15**: 47 (1961); G. J. Janz and D. W. James, *J. Chem. Phys.* **35**: 739 (1961).
7. G. J. Janz and S. C. Wait, Jr., in: H. A. Szymanski, *Raman Spectroscopy*, Plenum Press, New York (1967), Vol. 1, Chapt. 5.
8. T. F. Young and R. P. Westerdahl, *U.S.A.F. Office of Aerospace Research*, Contract No. AF33(616)-5697, ARL Report No. 135(1961); R. P. Westerdahl, Thesis, University of Chicago, 1961.
9. W. Bues, *Z. Anorg. Allgem. Chem.* **279**: 104 (1955).
10. J. H. R. Clarke, C. Solomons, and K. Balasubrahmanyam, *Rev. Sci. Instr.* **38**: 655 (1967).

11. J. H. R. Clarke and R. E. Hester, *J. Chem. Phys.*: in press (1969).
12. L. S. Arighi and M. V. Evans, *Appl. Spectry.* **21**: 43 (1967).
13. C. Solomons, J. H. R. Clarke, and J. O'M. Bockris, *J. Chem. Phys.* **49**: 445 (1968).
14. J. P. Young, *Anal. Chem.* **36**: 390 (1964).
15. A. J. Melveger, R. K. Khanna, B. R. Guscott, and E. R. Lippincott, *Inorg. Chem.* **7**: 1630 (1968).
16. J. R. Ferraro, J. S. Ziomek, and K. Puckett, *Rev. Sci. Instr.* **35**: 754 (1964).
17. J. E. Griffiths, R. P. Carter, and R. R. Holmes, *J. Chem. Phys.* **41**: 863 (1964).
18. N. Craig and J. Overend, *Spectrochim. Acta* **20**: 1561 (1964).
19. J. S. Kittelberger and D. F. Hornig, *J. Chem. Phys.* **46**: 3099 (1967).
20. S. M. Blumenfeld and H. Fast, *Spectrochim. Acta* **24A**: 1449 (1968).
21. J. I. Bryant, *Spectrochim. Acta* **24A**: 9 (1968).
22. R. Savoi and A. Anderson, *J. Opt. Soc. Am.* **55**: 133 (1965).
23. S. S. Bhatnagar, E. J. Allin, and H. L. Welsh, *Can. J. Phys.* **40**: 9 (1962).
24. V. Soots, E. J. Allin, and H. L. Welsh, *Can. J. Phys.* **43**: 1985 (1965).
25. A. H. McKague Roseveer, G. Whiting, and E. J. Allin, *Can. J. Phys.* **45**: 3589 (1968).
26. G. L. Carlson, *Spectrochim. Acta* **24A**: 1519 (1968).
27. G. L. Carlson and W. G. Fateley (personal communication).
28. J. H. Carpenter, W. J. Jones, R. W. Jotham, and L. H. Long, *Chem. Commun.*: 881 (1968).
29. A. R. Gee and D. C. O'Shea, *Rev. Sci. Instr.* **37**: 670 (1966).
29a. A. R. Gee and G. W. Robinson, *J. Chem. Phys.* **46**: 4847 (1967).
30. J. A. Koningstein, in: H. A. Szymanski, *Raman Spectroscopy*, Plenum Press, New York (1967), Vol. 1, Chapt. 3.
31. R. E. Hester and C. W. J. Scaife, *J. Chem. Phys.* **47**: 5253 (1967).
32. H. Brintzinger and R. E. Hester, *Inorg. Chem.* **5**: 980 (1966).
33. R. E. Hester and W. E. L. Grossman, *Inorg. Chem.* **5**: 1308 (1966).
34. A. Ferrari, A. Briabanti, A. M. Manotti Lanfredi, and A. Tiripicchio, *Acta Cryst.* **22**: 240 (1967).
35. R. E. Hester and K. Krishnan (unpublished work).
36. R. E. Hester and K. Krishnan, *J. Chem. Phys.* **47**: 1747 (1967); **46**: 3405 (1967).
37. R. E. Hester and K. Krishnan, *J. Chem. Soc. (A)*: 1955 (1968).
38. D. W. James and W. H. Leong, *Chem. Commun.*: 1415 (1968).
39. K. Williamson, P. Li, and J. P. Devlin, *J. Chem. Phys.* **48**: 3891 (1968).
40. R. E. Hester and K. Krishnan, *J. Chem. Phys.* **49**: 4356 (1968).
41. J. H. R. Clarke and C. Solomons, *J. Chem. Phys.* **47**: 1823 (1967).
42. J. H. R. Clarke and R. E. Hester, *J. Chem. Phys.* **50**: 3106 (1969).
43. L. A. Woodward and M. J. Taylor, *J. Chem. Soc.*: 4473 (1960).
44. J. T. Kenney and F. X. Powell, *J. Phys. Chem.* **72**: 3094 (1968).
45. J. H. R. Clarke and R. E. Hester, *Chem. Commun.*: 1072 (1968).
46. J. H. R. Clarke and R. E. Hester, *Inorg. Chem.*: in press (1969).
47. J. H. R. Clarke and C. Solomons, *J. Chem. Phys.* **48**: 528 (1968).
48. G. J. Janz and D. W. James, *J. Chem. Phys.* **38**: 902 (1963).
49. W. Klemperer, *J. Chem. Phys.* **25**: 1066 (1956).
50. J. J. Barrett and N. I. Adams, III, *J. Opt. Soc. Am.* **58**: 311 (1968).
51. I. W. Levin (personal communication).
52. F. Perzl and H. Moser, *J. Mol. Spectry.* **26**: 237 (1968).
53. R. E. Hester, in: H. A. Szymanski, *Raman Spectroscopy*, Plenum Press, New York (1967), Vol. 1, Chapt. 4.
54. R. Savoi and A. Anderson, *J. Chem. Phys.* **44**: 548 (1966).
55. J. E. Griffiths, R. P. Carter, and R. R. Holmes, *J. Chem. Phys.* **41**: 863 (1964).
56. E. L. Gasner and H. H. Claasen, *Inorg. Chem.* **6**: 1937 (1967).

57. J. H. Carpenter, W. J. Jones, R. W. Jotham, and L. H. Long, *Chem. Commun.*: 881 (1968).
58. W. D. Johnston, Jr., and I. P. Kaminow, *Phys. Rev.* **168**: 1045 (1968).
59. P. A. Fleury, S. P. S. Porto, L. E. Cheesman, and H. J. Guggenheim, *Phys. Rev. Letters* **17**: 84 (1966).
60. P. A. Fleury, S. P. S. Porto, and R. Loudon, *Phys. Rev. Letters* **18**: 658 (1967).
61. C. H. Henry, J. J. Hopfield, and L. C. Luther, *Phys. Rev. Letters* **17**: 1178 (1966).
62. J. K. Wilmshurst, *J. Chem. Phys.* **39**: 1779 (1963).
63. G. A. Jeffrey and M. Vlasse, *Inorg. Chem.* **6**: 396 (1967).
64. Y. Mikawa, R. J. Jakobsen, and J. W. Brasch, *J. Chem. Phys.* **45**: 4528 (1966).
65. T. Shiamanouchi, M. Tsuboi, and T. Miyazawa, *J. Chem. Phys.* **35**: 1597 (1961).
66. S. S. Mitra, *Solid State Phys.* **13**: 1 (1962).
67. S. Bhagavantum and T. Venkatarayudu, *Theory of Groups and Its Application to Physical Problems*, 3rd. ed., Andhra University, Waltair (1962).

Chapter 6

Raman Spectroscopy with Poor Scatterers

E. Steger

Sektion Chemie
Technische Universität
Dresden, DDR

DEFINITION OF POORLY SCATTERING SAMPLES

Any molecule or crystal will have vibrations of very different intensity in its Raman spectrum. By considering together equations (2), (19), and (27) from the introductory chapter in Volume 1, the intensity of the fundamental with number i ($i = 1, \ldots, 3N - 6$ or $3N - 5$, etc.), characterized by normal coordinate Q_i, may be written after introduction of Placzeck's simplified polarizability theory as

$$I_i \propto \left(\frac{\partial \alpha}{\partial Q_i} \right)_0^2 \tag{1}$$

The occurrence of Q indicates dependence on the mode of molecular motion. For symmetric molecules there are selection rules by which, in many cases, some of the vibrational intensities I_i are zero, since changes of polarizability in one part of the molecule are cancelled by opposite changes in others. Intensities may by indefinitely near zero for related molecules lacking such symmetry but retaining comparable vibrations. There may also occur low-intensity vibrations without any selection rules pointing to this possibility. For the molecule P_4O_6, one of the two totally symmetric vibrations was unknown until 1965, and the right frequency of the analogous movement of P_4O_{10} was recognized in 1967.[1] One out of the four totally symmetric vibrations of $P_3N_3Cl_6$ was discovered[2] and recognized[3] only after a series of misleading attempts.

If all this should not apply to a certain molecule under consideration, there will always be the harmonics. They have very low intensity in Raman spectroscopy (Volume 1, Chapt. 1, p. 31) if there is not the particular effect of Fermi resonance (Volume 1, Chapt. 1, p. 30). Until now, overtones and combinations in Raman spectra were only observed in rare or dubious cases.

Leaving those universal possibilities of occurrence of weak Raman lines beside stronger ones, we have to think of certain molecules or classes of compounds which give Raman spectra with only low overall intensity. This low intensity does not follow from a particular form of vibration [Q_i in equation (1)], but from small values of the molecular polarizability α. We may assume that with small α also the changes $\partial\alpha/\partial Q$ may only be small. It is not intended here to repeat Volume 1, Chapter 2 or this volume's Appendix. But measurements and theory treated there must be generalized to give some rules here of when a poor scattering power is likely to occur: there is high polarizability with atoms rich in electrons, so we have to expect on the other side low intensities for hydrogen motions, and Raman spectra of compounds from the Li–F row of the periodic table should be weaker than from those containing heavier elements. But the high intensity associated with π electrons is often predominant, and the statement concerning the lighter elements pertains to cases without any multiple bonds. Insofar as the heavier elements are metals, ionic bonds will predominate in their compounds. With increase of ionic character Raman intensities diminish. Scattering may drop beyond present possibilities of detection if highly ionic bonds are developed. But theory admits Raman lines for assumed purely ionic compounds as soon as mutual ion polarization is included, and, therefore, in many crystals considered generally as highly ionic Raman scattering has been detected. The chemical grouping most adverse to Raman investigation then seems to be the proton in a symmetric hydrogen bond.

Metal–carbon bonds have quite different degrees of polarity and, following from this, the Raman intensities also vary. The metal–metal bond has high scattering power.[4,5] Good scattering power is also found for semiconductors. The Raman intensity from GaP was estimated to be more than 10 times higher than from diamond.[6] A comparative measurement revealed a 1000-fold higher scattering cross section than for CdS.[7]

When we turned to simplified polarizability theory as a base for discussing Raman intensities, we lost the virtual connection to the electronic states of the molecule under consideration. It does not seem fortuitous that classes of compounds with low Raman intensities are also classes without UV absorption above 200 nm. This may not mean anything different than was already said in terms of polarizability theory, at least concerning the higher contributions from π electrons. But there are suggestions that the possibility of electronic excitation at longer wavelengths produces an increment of scattering power not contained explicitly in polarizability theory (see Volume 1, p. 838). We believe that the much lower Raman scattering from phosphoric acid and its derivatives, in contrast to comparable sulfur compounds, can not be understood from differences in the number of double bonds and bond polarity alone, because Chantry and Plane (for references see Volume 1, Chapter 4) by such suppositions reached conclusions on bond orders differing from those derived from vibrational frequencies and force-constant calculations.[8] Therefore, lack of long-wave UV absorption should be included as an independent feature signaling low scattering power.

To the compounds classified as poor scatterers by the rules given belong aliphatic hydrocarbons and phosphoric acid esters; they are not regarded as particularly difficult for Raman work because they are generally investigated as pure liquids. But limitations are encountered in studies with smaller concentrations. Anyway, we should include under the heading of poor scattering samples preparations of molecules (or such molecular species as ions, associates, etc.) which may only be encountered or must be investigated for special purposes as a gas or in solution.

Low concentration in solution presents difficulties which might be underestimated by infrared spectroscopists. The dependence of scattering on concentration is not logarithmic as it is for extinction in absorption experiments. Simple proportionality will not favor the appearance of those constituents that are less concentrated. In a radiation-scattering experiment, furthermore, the possibility of compensating for weakness of effect by something equivalent to extended optical path of absorption is more limited. Big scattering volumes are used in Raman spectroscopy of gases. Small sample size causes loss of energy in experiments with liquids, but samples of several cc still contain much more chemical compound than is needed for an infrared spectrum.

Lack of compound is especially aggravating the situation for poor scatterers. The question of detectability of weak lines must be considered from the relationship of Raman intensity to plate background, or of signal-to-noise ratio. All the different devices and measures for cutting down sample size tend to bring in more stray radiation from primary light. It was an early experiment[9] which showed that by putting the front plate of a narrow sample tube immediately up to the slit of the spectrograph (with adequate exciting lamps) there will be greater intensity and more background than with projection according to Nielsen.[10] Background and the gain in intensity originate from reflection from the walls of the tube in this case. More recent apparatus with antireflection coatings on lenses and beam splitters avoids energy losses and minimizes stray light, which is also reduced by double monochromators, but there is at least one published example pointing out that by employing larger-diameter tubes results with even the Cary model 81 spectrometer could be improved.[11] The limitations that are still apparent today with stray light, and occasionally with fluorescence caused by impurities, will warrant a special section on sample treatment in connection with Raman spectroscopy of poor scattering samples. The problems arising from the physical nature of samples will be considered here only insofar as they promote and aggravate the difficulties presented by certain classes of compounds because of their chemical composition or structure. These are the poor scatterers in our narrower sense. We will illustrate with some examples the difficulties in Raman investigations in such cases.

It is not the intention of this chapter to review all published work on certain classes of compounds from a certain period. Examples are only cited to prove expressed opinions and are selected so as not to duplicate cases unnecessarily.

The situation prevailing now (1968) in the field of Raman work with poorly scattering samples will be illustrated by some examples taken from the recent literature. To show the difficulties with first-row elements in compounds without multiple bonding, we look at work on bis(trifluoromethyl)peroxide.[12] The infrared spectrum was recorded. Raman data were needed to reach some conclusions about the structure of the molecule. The recordings had to be taken from the liquid at $-80°C$. Experienced workers with the Cary model 81 at their disposal must finally conclude that "unfortunately too few lines were recorded" to allow a complete comparison with the infrared data. The recordings were successful, however, in that many infrared active fundamentals were found in the incomplete Raman spectrum; symmetry C_{2h}, which means mutual exclusion, was ruled out. There were also nine polarized bands observed, and this is more than the number possible for symmetry C_{2v}. The question whether point group C_2 or simply C_1 should be accepted remains unsettled.

Similar results were reached with phosphine boran, BH_3-PH_3.[13] The 11 fundamentals of infrared absorption were found in the Raman spectrum, but in four places this band was so very weak that the frequency could not be measured. Without previous knowledge of the infrared spectrum in connection with the special selection rules, which only allow coincidences, these bands probably would have gone undetected. The investigation was done with photographic equipment on the liquid and the solid. Exposure times were up to 72 h on Kodak IIa-O plates with excitation by General Electric AH-4 lamps.

Some examples of the difficulties with highly ionic compounds containing polarizable atoms should also be included. An investigation with the Cary model 81 laser Raman spectrometer of the insoluble polymeric chromium(III) alkoxides [14] brought out only four bands (eventually with shoulders) between 160 and 480 cm^{-1}; this should be compared to a sufficiently great number of infrared absorptions, mostly at higher wave numbers.

Sodium uranyl triperoxide $Na_4[UO_2(O_2)_3] \cdot 9H_2O$,[15] investigated with the Perkin–Elmer LR-1 (with Spectra-Physics He–Ne laser 125), yielded four bands and two additional shoulders from the 10 Raman-active fundamentals expected from the selection rules of symmetry D_{3h} for the anion, when investigated in the solid state. An aqueous solution containing $Li_4[UO_2(O_2)_3]$ yielded two Raman bands only, one polarized and the other depolarized.

In photographic work with the Hilger E612 Raman spectrograph, knowledge of the active fundamentals of the ions SbF_6^- and AsF_6^- was obtained from the solutions of the potassium salts in N,N-dimethyl formamide (4.4M) and acetonitrile (1.75M), respectively.[16] In melts from $SF_4 \cdot SbF_4$, $SeF_4 \cdot SbF_5$, $TeF_4 \cdot SbF_5$, and $SeF_4 \cdot AsF_5$ it was difficult to identify and understand the vibrations of the anions and cations. Polarization measurements would have been helpful, but the slow decomposition of the melts did not allow the long exposure times needed.

High bond polarity and small atomic polarizability occurring simultaneously may especially lead to disappointment, and this must be expected with hydrogen bonds. To study such Raman bands some tetraalkylammonium hydrohalides, $[R_4N]ClHCl$,[17] were investigated in the solid state and in solution with the Hilger E612 having an interference filter arrangement devised by Brandmüller for solids.[18] The spectra were of good quality and showed the cation bands clearly, but only in the single case of solid $[(CH_3)_4N]ClHCl$ was it possible to discern an anion band at 1180 $\pm$ 10 cm^{-1}, broad and with very low intensity. These efforts were continued with hydrobromides. Two of the three vibrations of linear $BrHBr$[19] known from infrared absorption were discovered in the Raman spectra.

A similar problem is encountered in the protonation of ethers. This was tackled now for the case of mixtures with anhydrous hydrogen fluoride,[20] showing no Raman lines but some continuum. (The fluorescence generally occurring with samples of commercial hydrogen fluoride was reduced with good success.) The Raman bands from $(C_2H_5)_2O$ and C_2H_5OH were observed, and new bands appeared signifying the formation of a new molecular species at higher hydrogen fluoride concentration; none of these bands could be attributed to a motion in an $O \cdots H \cdots F$ or an $> O \cdots H$ group. This work was done with the Cary model 81 spectrometer. The sample tubes were made of polychlortrifluorethylene with a sapphire front plate.[21]

With single crystals, which offer the best opportunities for observation of OH modes, the situation is not very much better. Working with the Cary model 81, mercury arc excitations on a laboratory-grown crystal of teepleite, $Na_2[B(OH)_4]Cl$,[22] disclosed only two of the OH stretching vibrations expected from crystal symmetry and only one from the two in-plane bending vibrations. OH out-of-plane bending, known from the infrared to be at $614\,cm^{-1}$, was not detected. No measurements with polarized radiation were attempted because of the too low intensity of all the lines.

The not always recognized low scattering power of phosphates may be noticed in O^{16}-enriched samples of K_3PO_4 in aqueous solution.[23] Raman spectra were recorded with a Cary model 81 instrument (mercury arc excitation); very interesting isotopic shifts were discovered, but only those arising from the totally symmetric vibration at $937\,cm^{-1}$ were investigated. Other bands did not appear in the recorded spectra of the solution, "due to the extreme weakness of these bands, which could therefore be observed only spectrographically."[23]

INSTRUMENTATION CONSIDERATIONS

Photoelectric Recording vs Photographic Plates

Formerly it was necessary to put primary emphasis on the preparation of the sample. This situation is about to change by the introduction of high-aperture double monochromators and laser excitation into Raman spectroscopy. We should start discussing, therefore, the progress made possible by the achievements in instrumentation.

One historical milestone in Raman spectroscopy is photoelectric recording (*cf.* Volume 1, Chapter 2). This technique should not, however, be overemphasized here. Due to experience with poor scatterers, photoelectric recording was looked at with distrust by some workers, and the possibility of making sure of the completeness of a Raman spectrum seemed to get lost. It was held that by prolonging the exposure time until there were no new bands (even with very long-lasting experiments) the completeness of data could be proven. This principle does not work well in every case, e.g., the questionable structure work on trisilyl phosphine, $P(SiH_3)_3$, done recently by photographic techniques.[24] Raman spectra had been obtained from the liquid at $-60°C$ with a Hilger E612 spectrograph fitted with a Toronto arc source. Since only one polarized line with no coincidence in infrared absorption was found as a possible skeletal vibration, it was concluded that molecular symmetry D_{3h} with a planar arrangement of P and Si atoms, as known from $N(SiH_3)_3$, must be present. Microwave[25] and P^{31} nuclear resonance work,[26] however, led to a pyramidal structure. The Raman spectrum has been taken as complete since the prolonged exposure time of 30 to 90 min did not bring any new bands, and the spectra were of good quality. It seems that for such a poor scattering compound reliance on the prolongation principle may not be justified. More powerful apparatus would be desirable, but to improve on the results obtained with poor scatterers by patient and accurate photographic work, investment in the most powerful apparatus available would be imperative.

At first the application of photomultiplier tubes was irrelevant to progress with poor scattering samples. The truth of this may be illustrated by the history of research in water association. It was discovered in 1932[27] that there are some Raman bands in the spectrum of the liquid, which are recorded now at $60\,cm^{-1}$, $175\,cm^{-1}$, $450\,cm^{-1}$, and $780\,cm^{-1}$ (they are somewhat diffuse).[11] Magat, in 1934, confirmed these first observations, and explained the bands as belonging to intermolecular vibrations.[28] He also found the corresponding bands, except of the most low lying, in D_2O.[29] The measurements were repeated and completed in 1937[30] and 1948.[31] When Hornig[32,33,34] turned to the investigation of water association by means of photoelectrically recorded Raman spectra, he and his coworkers only considered the stretching and bending regions. Apparently the homemade spectrometer[32] failed to respond to the quite weak librational bands. (It is generally taught that spectrometers of the Littrow type will especially suffer from stray light.[35]) Walrafen achieved the registration with the Cary model 81 instrument in 1962,[36] and later managed an even more rigorous exclusion of stray light.[11] Single-slit operation was used, as well as 19-mm-o.d. tubes masked on the outer surface near the ends. Commonly observed mock lines from the mercury arc around $175 \, cm^{-1}$ were avoided by these measures; but it was still impossible to obtain a degree of depolarization for the band at $60 \, cm^{-1}$ because of its low intensity and the vicinity of the exciting line.

Photoelectric recording in the water example was important to establish in a quantitative manner the intensity changes with higher temperature, which were noted from the beginning.[27] There are further cases with difficult samples where vital information on band contours was obtained. For the zinc chloride melt which had been investigated twice before, photoelectric recording[37] disclosed that the $230\text{-}cm^{-1}$ band is asymmetrical and consists of at least two closely spaced lines not previously reported. The spectrometer in this work was a three-prism Steinheil spectrograph. The photomultiplier tube was moved across the focal plane. Four inverted U-tube mercury lamps with water-cooled electrodes surrounded the sample.[38] Even if highly ionic, however, zinc chloride has good scattering power, at least for the more symmetrical vibrations. For this reason also sulfuric acid solutions, investigated with the same apparatus,[38] do not belong to the very poor scattering samples. With similar equipment Soviet workers found formic acid dimers $(HCOOH)_2$ in aqueous solution[39] (the three-prism spectrograph ISP-67[40] was used, with accessories for photoelectric recording).

Stamm and coworkers[41] published information on relationships between photographic and photoelectric work. They found that with their instrument (see Volume 1, Chapter 2) they could record those lines which appeared in a 2-h exposure. With the Hilger E612 spectrometer, exactly the same estimate was reached in the course of experiments in Dresden. Only the strongest lines (if any) are photographed from poor scatterers in 2 h.

The final stage of photoelectric recording is (as already reached in infrared spectroscopy) digitized output, which may be fed to computers. This prospect is of interest in Raman spectroscopy of poor scattering samples, since better signal-to-noise ratio and subtraction of background may be provided. Raman bands which are not detected with a given source and resolution should become measurable. Ziegler and Hoffmann[42] have reported on such studies. They transferred the intensity data from about 1000 points of a Raman spectrum to a time-averaging computer in 10 repetitive runs, and obtained remarkable reduction of noise. Subtraction of a solvent spectrum was performed step by step until a selected band of the solvent, isolated from solute bands and not affected by raising or falling of background, was reduced to zero. If background is to be subtracted, it should be measured without changes in sample or alteration of optical alignment. This is possible just by opening the slit to such an extent that all the Raman lines disappear. Intensity distribution in the continuum will not depend on slit width.

High-Aperture Double Monochromators

For many years only two complete commercial Raman spectrometers of this type existed: the Cary model 81 (since 1953; see Volume 1, Chapter 2) and the $\Delta\Phi$C-12 (known outside the Soviet Union since 1958) treated by Brandmüller and Moser[48] and (more fully) by Brandmüller and Moser.[40] Table 1 in Volume 1, Chapter 2 of this work shows that the double monochromators are really a new class of instruments. Apparatus of this type facilitated work on many compounds which had formerly been troublesome to investigators. Polycrystalline solids may often be investigated in the same way as liquids and only with the high-aperture double monochromators did the benefits of photoelectric recording become attainable for poor scattering compounds. Better results on band intensities, band shape, and, therefore, true band maximum wave numbers, and, particularly, better degrees of depolarization, are now also obtained for poorly scattering samples. In many cases a more final state of knowledge seems to have been obtained. Photoelectrically recorded data are now looked on as being more reliable, even if the good photographic work of the past is not exceeded very much. This latter disenchanting aspect is also brought to attention when recent literature on poor scatterers is reviewed.

Gerding and his coworkers have achieved admirable results with samples which are difficult to handle and poorly scattering. Investigations on melts of $AlCl_3 \cdot NaCl$,[44] which are supposed to contain the tetrahedral ions $AlCl_4^-$, were now performed with the Cary model 81 instrument.[45] Almost the same frequencies and properties of polarization were found for the real bands and two others, which had been given dubious stature, were now disproved. Perhaps they had originally been mistakenly seen in the strong background of the photographic plates. The $AlCl_3 \cdot KCl$ melt does show additional bands in photo-electric recorded spectra.

Ionic melts are treated elsewhere in both volumes of this book, so we will not give further examples of this kind here. The existence of these chapters shows that much progress has been achieved in the past years in this field.

The Raman spectrum of hydrazine, N_2H_4, had been studied six times, including one investigation with a Cary model 81 instrument,[46] when Durig and coworkers recorded it again.[47] They accepted the data given by Ziomek and Zeidler,[46] considering them the most reliable; but these were the wave numbers obtained by Goubeau in 1940,[48] with only (still incomplete) data on polarization added. Equilibria between molecules of different structure and association are the big problems with hydrazine. Raman spectra at elevated temperatures and at different concentrations in dimethylsulfoxide were recorded,[47] as well as spectra from 20 vol.% of N_2D_4 in D_2O. Spectral slit width was 10 cm^{-1}, with remarkably low noise. Work of this type and quality could not have been done on photographic plates. The similar molecule N_2F_4 seems to possess even less scattering power. Kotov and Tatevski[49] succeeded in photographing the Raman spectrum of the gas with a prism instrument by choosing the short focal length of 12 cm and spectral slit width of 16 cm^{-1}. Work with the Cary model 81[50] with the liquid at -150 and $-80°C$ resulted in further resolution of some of the bands and brought out four more at small frequencies. Measurements of degrees of depolarization were accomplished, also. It was shown that in addition to the dominating gauche form of C_2 symmetry there exists the *trans* or C_{2h} form in every state and every liquid temperature. Noise amounts to 10% of the strongest line in the Cary work, however, and if the parameters of recording chosen for the published spectra represent the highest grade of instrument performance, then it is not possible to record any bands from a 10% solution.

With special reference to the poor scatterers, there are many examples in the literature where limitations of the double-grating instruments with mercury arc excitation become apparent. Difficulties with mock mercury lines around 175 cm^{-1} were mentioned in connection with the investigation of liquid water[11] earlier in this chapter. At higher wave numbers similar difficulties may arise, such as experienced with hydrogen fluoride in the solid state.[51] Background scattering around the 4916-Å mercury line, which corresponds to a 2602-cm^{-1} Raman shift, obscures the region of the valency vibration of the DF molecule or rather the multiplet expected from crystal field splitting. By subtracting the background, which was taken from the scattering of a sample of solid HF, and multiplying with a scaling factor, a corrected curve was obtained with only one small peak at 2524 cm^{-1}. In this paper[51] figures showing smoothed curves are given, but certain portions are drawn including the noise experienced. This helps greatly to attain realistic judgement. Also, in the region of the lattice vibrations the spectra revealed only one line each for HF and DF (at 550 cm^{-1} and 398 cm^{-1}, respectively) at the very limit of instrumental performance, as was stated. The band observed in each case could be the out-of-plane libration of the zigzag chain unit. Until further success in Raman spectroscopy is achieved, there remain different possibilities in assigning the in-plane librations to the observed infrared bands. Before the theme of too low excitational intensity is taken up, comprehensive information on possibly occurring mercury lines seems desirable.

In the part of the spectrum up to a 1500-cm^{-1} Raman shift from 4358 Å we noticed a number of mercury lines with very long photographic exposure times[52] which do not seem to have been reported before. They are given in Table 1. For the longer wavelengths we have nothing to add to tables which are to be found in the literature[53] or are issued with the Cary model 81 spectrometer.

After the mercury lines we must consider interference from the Raman spectrum of the glass from which the sample tubes are made. In our investigations[52] we observed within the Raman spectra of aqueous solutions or solid samples the Raman lines of glass recorded in Table 2. The intensities are estimates from photographic plates, given relative to the strongest band of glass at 1233 cm^{-1}. The glass used for our sample tubes is of the borosilicate type (Jenaer Geräteglas 20: 74.5 SiO_2, 4.6 B_2O_3, 7.7 Na_2O, 0.8 CaO, 3.9 BaO, 8.5 Al_2O_3). The same lines had been observed from a solid rod of glass 30 years ago with differences of ± 2 cm^{-1} for the weakest lines.[57] Only the stronger lines are encountered in observations on liquid samples. In the experimental arrangement no projection of the sample volume into the collimator was used; the front plate of the tube was set near the slit, so that radiation generated in the walls could enter the spectrograph—directly or after reflections—to a higher extent than this is possible with projection.

Table 1. Mercury Lines between 0 and 1500 cm^{-1} Raman Shift from 4358 Å Encountered in Raman Spectra

Submersed lamp,[54] solid samples[52]	Water-cooled electrodes,[55] liquid samples[52]	Toronto arc (Cary model 81)	High-pressure arc[56]
994 ($\frac{1}{2}$)			
833 (1)			
656 (1)	656 (1)		
482 (3)	482 (3)		481 (1)
307 (1)			308 (1)
169 (10)	169 (10)	169	
144 (3)	144 (3)	145	
105 ($\frac{1}{2}$)			148
92 (2)	92 (2)		to
63 (2)	63 (2)		93 : continuum

Table 2. Raman Lines of Glass Actually Observed in Raman Spectra

Liquid samples, cm^{-1}	Solid samples, cm^{-1}	Relative intensity, $I\,(1233) = 10$
1384	1384	8
1233	1233	10
	1122	1
	1028	2
	908	1
809	809	3
711	711	2
	577	5

The weaker lines were observed in an arrangement for poly-crystalline solids, where the sample is put into a tube which has a concave cone instead of a front plate. The walls of this cone are illuminated to the same extent as the sample. Since the scattering power of phosphates and salts of related acids is comparable to that of glass, Raman bands are to be expected in investigations such as this.

It must be pointed out that the powerful mercury arcs and the most ingenious optics of commercial Raman spectrometers are still inadequate to many problems. To illustrate the difficulties possibly encountered with highly ionic compounds, even from fairly polarizable atoms, an investigation with the Cary 81 on $(NH_4)_2SbCl_6$ may be cited.[58] It was not possible to obtain spectra from solutions, and from those of the polycrystalline solid of the species b_1 and b_2 of the

C_{4v} pyramid, which are inactive in the infrared, there was observed beyond doubt a v_4 with medium intensity at $420 \, cm^{-1}$. Vibrations v_5 and v_6 are still missing or dubious, respectively.

For compounds $(CH_3)_2PSBr$, $(CH_3)_2PSCl$, and $(CH_3)_2POCl$ agreeable results were obtained[59] with a Cary 81 spectrometer with mercury arc excitation on benzene solutions. The authors stress, however, that owing to the weakness of the spectrum in the case of $(CH_3)_2POCl$, polarization data on this molecule are only qualitative. For the spectrum of the crystalline solid the authors took recourse to a Cary 81 instrument equipped with a He–Ne laser source.

There remains, in the case of insufficient Raman intensity, the trivial possibility of sacrificing dispersion by working with very short focal lengths and broad slits. An extreme example is provided by Hirano.[60] When every attempt to observe a Raman line from DNA (solution of 2 mg calf-thymus DNA in a 25-ml buffer solution) with a commercial Raman spectrometer had failed, a photoelectric spectral fluorometer was employed. Excitation at 260 μm and 402 μm revealed scattering with three maxima and one additional shoulder. These were partially assigned to PO^- or purine ring vibrations. At the same time, Raman spectra of quite good resolution were obtained[61] from aqueous solutions of the purine bases (maximum concentration, $1M$). These had formerly been studied in solution only by infrared absorption in $CHCl_3$ and CCl_4, and in these solvents they exhibited hydrogen bonding. There is no sign of this or any base pairing in aqueous solution, as shown by the Raman studies cited; apparently there are still other requirements for the stabilization of the double helical structure.

Evidently, for instruments with mercury arc excitation, samples free from fluorescence are still required. In an examination of melamine, $N_3C_3(NH_2)_3$,[62] with a Cary 81 instrument it was noted that fluorescence could not be eliminated by recrystallizations. Because of unfavorable solubility the investigation could proceed only with the solid. No measurements of degrees of depolarization were possible and nothing could be detected from angle deformation or rocking vibrations of the NH_2 groups. In spite of these limitations the information obtained led to a satisfactory assignment in many cases where that was not previously possible in six different investigations of the infrared spectrum alone.

The examples treated should show that with mercury arc sources the high-aperture double-grating monochromators still suffer from mock mercury lines from the source still not being powerful enough, and from fluorescence. These disadvantages are emphasized with commercial instruments, because the user is committed to the single wavelength of 4358 Å of the mercury arc.

It was considered, at one time, an error in methodology if the Raman spectrum excited by 4358 Å was not confirmed by further exposures with the 4047-Å exciting line, although this is not a requirement today. The interfering mercury lines may be singled out, since photoelectric recording presents the possibility of subtracting background. Furthermore, the old problem of excitation by lines at 4348 Å, and even 4339 Å for the strongest bands, is almost nonexistent now since the mercury arc with water-cooled electrodes shows these lines, originating from higher excited electronic levels, only weakly. The principle, therefore, that each spectrum should be obtained from two independent experiments has become unimportant. But it is not meaningless.

Being locked to one single excitation line also deprives one of the possibility of coping with the different problems of fluorescence and insufficient excitation energy by changing to other exciting lines.

Fluorescence (and absorption by colored samples) may be principally avoided by excitation at longer wavelengths, and weakness of scattering may be cured by short-wave excitation. Work with colored and fluorescing samples will not be treated in this chapter although insofar as fluorescent impurities often set a limit to investigation of poor scattering samples we will discuss possibilities of avoiding these limitations. Rigorous exclusion of the 4047-Å mercury line is often sufficient. This is best done by means of a filtering solution of $NaNO_2$ or, better, KNO_2 which is the more soluble. Lines of longer wavelength will seldom excite fluorescence (there are copious statements in the literature concerning this, e.g., in a recent paper on aqueous solutions of zinc chloride[63] we may read that all efforts with recrystallizations were in vain, but with nitrite as an optical filter it was possible to record spectra at first with a prism spectrometer of 12-cm^{-1} slit width and then with a double monochromator of 4-cm^{-1} slit width).

The value of excitation at short wavelengths may be seen from the work on liquid water cited earlier. The early discovery[27] of intermolecular vibrations was due to excitation with the mercury line at 2537 Å. Magat[28] obtained spectra excited by the 4358-Å line only with much effort, and Hibben[30] employed the ultraviolet line again.

The 2537-Å mercury line was already important in the first years of Raman spectroscopy since it made possible early work on more difficult solids[64] and the first observations of pure rotational Raman spectra.[65] This was not only possible because it provided the maximum excitational power, but also because it is a line originating from a first excited level above the ground state of the atom, and therefore it may be reabsorbed between the sample and the spectrograph. This latter possibility is of the same value as employment of a double mono-chromator, and for this reason, combined with its high excitational power, this line is still used in experiments of the type stated.[66] There were even investigations published recently aimed at reabsorption of the 4358-Å mercury line by a plasma adequately excited by irradiation with 2536 Å or by electron bombardment.[67]

Even use of the mercury line at 4047 Å for excitation may offer opportunities over the 4358-Å line. Better spectra from solutions of aliphatic amines[68] and from polycrystalline layers of methylamine CH_3NH_2 and the N deuterated species[69] were obtained by excitation with the violet line. (The observation by the authors that there was more intensity at 4047 Å in the arc seems peculiar. This line together with the one at 4358 Å and the green one at 5461 Å forms a triplet originating at one excitational level.) Exposure time with the solids was 60 to 70 h on Kodak 103aO plates; spectral slit width was 2.3 to $3 \, cm^{-1}$.

Lasers

Progress in the investigation of poorly scattering samples made possible by laser excitation may be found to a greater extent in work with solids. Since the highly developed technique for liquids provided satisfactory results for many compounds, the possibilities of laser excitation were concentrated at first on problems where either the greater wavelength, the high density of radiant energy in a small volume, the exactly traceable geometry of radiation paths, or the perfect polarization provided by the Brewster windows was of particular importance. The first laser Raman excitation studies were done with good scattering liquids, such as benzene or carbon tetrachloride. But there is some published work on more difficult samples even with the pulsed ruby laser, e.g., with an ionic melt[70]: the colorless, clear liquid which stannous chloride forms at 247°C has been investigated with the Cary model 81 spectrometer, mercury arc excitation. At somewhat higher temperatures the color of the liquid becomes yellow or even red. For the equimolecular mixture of $SnCl_2$ and KCl complete transformation to $KSnCl_3$ may be assumed. With a ruby laser a Raman spectrum was photographed within 10 to 20 sec by 5 or 10 flashes, and 100 flashes did not yield any greater detail. The spectrum of the pyramidal anion was the same as that known from solutions in ether,[71] except for broader bands.

When long-wavelength excitation proved its value here, the possibility of exciting spectra from very small volumes was also demonstrated with really poor scatterers: Raman spectra were obtained for 1,1-dimethyldiborane, trimethyldiborane, and tetramethyldiborane, $(CH_3)_2BH_2BH_2$, etc.[72] from the liquids at $-70°C$ with samples of only about $0.1\ cm^3$. They had been sealed into Pyrex capillary tubes of 1.5 mm bore at $-196°C$. The laser beam (Ar^+, 4880 Å) was focussed through the rounded, downward end of the tubes, and formed an illuminated cylindrical portion of about 0.1 mm diameter within the sample. The scattered light was projected with a high-aperture lens into the spectrometer slit. Twenty bands or more were recorded for each compound, including results on polarization. It has been mentioned that possibly the depolarized BHB vibrations are of such low intensity that they still may be missing. Here it seems as if photographic work would have led even further.

The ease with which the laser beam may be directed into cooled samples within a Dewar vessel will promote the study of the solid phases of compounds now investigated only as liquids or gases. Such work yields spectra of the lattice vibrations, which will disclose molecular motions and intermolecular forces in the solid state. Incidental to this, knowledge of phase transitions is obtained, e.g., for dioxane where an investigation with a He–Ne laser of 170-mW output is reported.[73] The spectra were recorded with the Coderg PH 1 double-grating monochromator. Six bands from 59 to 156 cm^{-1} were obtained for the one modification with known crystal symmetry. This may be understood by the hypothesis of two molecules in the unit cell, since only librational motions should be active in Raman scattering. These bands become steadily broader with rising temperature and are shifted, so that just below the transition point only three maxima are discernible. At 272.9°K the transformation to a new solid phase takes place, also with three maxima, but quite different lattice bands. Marked changes at this well known transition point have been noticed for the internal molecular vibrations in the infrared spectra,[74] but this has not been possible in the Raman spectrum between 200 and 3200 cm^{-1}.

Cyclohexane C_6H_{12} with approximately the same low scattering power as dioxane, was investigated similarly in a cold cell, but with an Ar^+ laser.[75] Much of the site symmetry splittings, which were expected by Ito,[76] are now resolved (doublets for original a species and quartets for original e_g are observed). Lattice modes were also discovered, and the improved knowledge on the internal vibrations of the molecule is remarkable. Similar progress may be in the offing in many other cases.

The difficulty to be overcome in the investigation of polycrystalline samples, and especially with fine-grained powders, is the scattering of the exciting light into the spectrometer by the reflections from the vast number of optical interfaces made up of all the grain boundaries. This difficulty is found enhanced when using Raman spectroscopy in the study of catalytic phenomena or, more generally, in questions of adsorption. To get a great amount of sorbate bonded to a substrate, the latter should have a particularly large inner surface, meaning a very extensive optical inhomogeneity. Double monochromators[77] and arrangements with filters between sample and spectrograph[78] have made initial attempts in this field possible. These were restricted, however, to good scattering molecules adsorbed to not too active substrates where only physical adsorption occurred; for example, with antimony trichloride, acetonitrile, and acetophenone on silica gel it was possible to record some bands with the Raman spectrometer $\Delta\Phi C$-12 and to show the differences between adsorption and capillary condensation.[77] In the first attempt with laser excitation in this field, Hendra and Loader studied only good scattering molecules on silica gel,[79] and observed scarcely any signs of interaction in the spectra. They did not find disturbance of the selection rules of the free molecules in the case of carbon disulfide or *trans*-dichlorethane. More recently,[80] they could show with the new species formed from acetaldehyde on active alumina that Raman spectroscopy in fact offers a greater prospect for such investigations than infrared absorption studies, which were almost exclusively done until now. Young and Sheppard[81] employing such methods had observed the formation of a new species in the case cited, but from the limited spectral range accessible outside the absorption of the alumina, they could only conclude that the aldehyde must have been transformed to a chemisorbed state or to a polymer. From the Raman spectrum it is to be seen clearly that the five bands recorded are just the stronger ones of paraldehyde, which apparently is formed by catalytic action. By this experiment it is shown that, in principle, Raman spectroscopy is the more powerful and more appropriate method for studying chemisorbed species.

Active substrates for chemisorption will be highly polar in structure, and therefore give strong infrared absorption, obscuring broad regions of the spectrum. But these substrates will show low scattering power in their Raman effect, which will not interfere with the spectra of even moderate scattering adsorbates.

A particular and important class of polycrystalline solids are high polymers, and since they only form small crystals or, as a bulk sample, small domains of crystallinity interspersed with amorphous ranges, they are especially difficult for Raman work. Some, in addition, belong to the poorly scattering compounds. For polyoxymethylene (hexagonal and orthorhombic modification, deuterated specimen) Raman data with laser excitation were obtained recently.[82] It is stated that the spectra were of very low intensity. Therefore, they were recorded with spectral slit widths of 5 to 10 cm^{-1} with the Cary model 81 spectrometer. The instrument used was limited in the red to record only to 2000-cm^{-1} Raman shift. The method is still far from ideal, but spectra are reported that are vastly superior to those obtained before. Many fundamentals seem to remain still unobserved because of low intensity or overlapping, but comparisons with calculations for the complicated helical structures could be made. This was especially important for the case of the hexagonal modification, since more than half of the fundamentals are forbidden in the infrared.

Experimental results more complete than for these powdered samples were obtained in an investigation of polyethylene in the form of stretched fibers.[83] By a draw ratio of 1:10 very good orientation of the crystallites is obtained. The fibers were assembled in a tight bundle about 5 mm in diameter, brought into the laser beam, and aligned, successively, with the fiber axes parallel or at right angles to the beam and its plane of polarization. Since there is no orientation around the fiber axes to be expected, spectra recorded from different orientations, in this sense, were extremely similar. The results yielded directions of the transition moments for vibrational excitation that are in agreement with those obtained earlier from infrared dichroitic measurements for most of the bands. But there had been disagreement among authors until now, some favoring an assignment v_{CC} $B_{2g} = 1133$ cm^{-1}, v_{CC} $A_g = 1065$ cm^{-1} [84,85] and some the opposite.[86,87] Only the given assignment[84,85] is in accordance with the different Raman intensities from different orthogonal directions.

Working with polyethylene crystallites oriented in fibers or foils by stretching is similar to working with single crystals. Knowledge of the directions of the transition moments of vibrational excitation is obtained by measuring the intensity of scattered radiation in the crystallographically different directions with different planes of polarization. We will be brief in this section, because this work is usually carried out in physical laboratories, and there is a chapter on laser Raman spectroscopy in this book written by a physicist. We must state here, however, that work of this kind has been extended to highly ionic crystals of very low scattering power.

With He–Ne laser excitation there were, for example, recorded the fundamentals from the unit cells of BeO[88] and ZnO,[89] and the hydrogen motions in KH_2PO_4.[90] The more powerful Ar^+ laser was employed for the study of Al_2O_3 (corundum or sapphire).[91] Its fundamentals are considered to be 1000 times weaker than those of diamond. It should not be forgotten that Krishnan in 1947[92] had already photographed some bands by excitation with the mercury line at 2537 Å. Perovskite-type crystals had been investigated with Hg 4358-Å excitation already ($SrTiO_3$ and $NaTaO_3$).[93] More recent work[94] with the Ar^+ laser is noteworthy since a crystal brownish in color which absorbed some laser radiation still yielded a Raman spectrum.

In these studies, beside the fundamentals, harmonics were also recorded, despite the weakness of Raman lines of the second order. The scattering probability for first overtones and binary combinations is considered to be only $1:10^9$ as compared with $1:10^6$ for fundamentals.[6] There are crystals where selection rules prohibit fundamentals from Raman scattering, and only a second-order Raman spectrum is allowed. MgO (periclase) is one example. It was investigated by means of a 30-mW He–Ne laser, but there remained some doubt, whether impurities had contributed to the spectrum obtained.[95] This work has been brought to a more final state by making use of an Ar^+ laser.[96] Only one fundamental, but an extended second-order spectrum, is exhibited by CaF_2 (fluorite) with He–Ne laser excitation.[97] Conditions are similar for ZnS (sphalerite). The second-order spectrum was recorded with Ar^+ laser excitation.[98]

For NaCl, KCl, etc., there is a second-order spectrum only, and this has been well known for many years.[99,100] Now the question attracting interest is what disturbance of the cubic symmetry could lift the prohibition of the fundamentals. First observations of this kind[101,102] were traced to the formation of F centers by irradiation from the mercury arc, and for mixed crystals KCl, KI (0.01M), or

KCl, LiCl (0.005M), no effects were discovered by investigation with Hg 2537-Å excitation.[103] But from crystals KCl, KBr ($<8\%$), the impurity-induced first-order spectrum was recorded by excitation with an Ar$^+$ laser.[104]

Powerful Ar$^+$ lasers also extend the possibilities for detection of species dissolved in crystals. Melts from RbCl and other alkali halogenides were doped with K$_2$O,[105] and the concentration of the O$_2^-$ ions, which had entered the crystals grown from these melts, was estimated by ESR. An Ar$^+$ laser of 500 mW at 4880 Å (Spectra-Physics model 140) excited the Raman band of the anion and permitted recording by a Spex double monochromator fitted with an EMI 6256S photomultiplier with photon counting detection down to concentrations as low as 1 ppm.

Our examples of achievements in laser Raman spectroscopy, together with others in the literature, will give the impression that any problem presented by poorly scattering samples can be solved by employing powerful laser sources in connection with double-grating monochromators. This belief seems justified, but there are still limitations discernible today. Since the purpose of this chapter is to help in judging what problems may be solved with what apparatus or, vice versa, what apparatus will be needed to solve a certain problem, we must also discuss disappointments with laser applications, as far as such instances may be collected from the literature.

The exploitation of laser excitation for Raman spectroscopy is still in its initial stage today (1968). In 1928, the first year after the Raman effect was discovered, there appeared approximately 180 papers on this new subject. Sending a laser beam into a sample should be considered a step comparable in experimental setup to putting a mercury lamp to the side of a tube filled with a liquid. But in the years following the first announcements on Raman spectra obtained by laser excitation in 1962, the number of laboratories from which published results appeared remained small. Powerful apparatus and some experience seems to be important for success with difficult samples, and some published work may be of a preliminary nature.

We should not regard potassium dichromate, really, as a poor scatterer, since its absorption in the visible range is near to the laser wavelength of 6328 Å; but it is, at least, a compound with highly ionic bonds. After Stammreich and coworkers had investigated solutions by excitation with helium lines,[106] their work was repeated in the laboratory of Lafont[107] with a 3-mW He–Ne laser, and recently in the

U.S. with a similar source of not much more energy (6 mW).[108] The latter investigation seems to have been more successful. While the French group did not reinvestigate the aqueous solution, only confirming all the Raman bands reported by Stammreich, there are now corrections in wave numbers given and one band is considered as nonexistent. Degrees of depolarization were also obtained only recently. From the crystal bands some are resolved into components, but only one maximum is found instead of the three given[107] at 700, 753, and 772 cm^{-1}. Bands given by Stammreich for infrared absorption of the crystal are now confirmed by Raman results, but the low-frequency bands given by the French group received corrections in number and wave numbers, and are not in agreement with longwave infrared measurements. This is enumerated here to illustrate what the technical progress of two years means in the field of laser-excited Raman spectra.

The quite significant, but still very limited success in the examples of chromium alkoxides[14] and uranyl peroxides[15] cited in the introductory section of this chapter, shows that fine powders or solids of low crystallinity, which was probably the nature of the samples investigated, are a difficult matter for the commercial laser Raman spectrometers. Absorption may add difficulties in the case of green compounds. With the He–Ne laser Cary model 81 instrument this was encountered also in an investigation on chromium oxide Cr_2O_3.[109] A Raman spectrum was obtained, but apparently at the limit of instrument performance. It was "very noisy (color problem)."

Laser Raman spectroscopy, therefore, is not only dependent on laser power, but also on the correct laser wavelength. This may be seen clearly from a case where single crystals of fluorosilicates, $[Zn(H_2O)_6]$-SiF_6, $[Fe(H_2O)_6]SiF_6$, and $[Ni(H_2O)_6]SiF_6$, were to be investigated.[110] The crystal of the zinc salt was studied by the 4047-Å emission of an Ar^+ laser, but the spectrum of the iron salt, which absorbs this radiation, was excited by the mercury line at 4358 Å. No Raman spectrum could be obtained from the crystal with nickel as the metal ion.

Not only is it a problem to couple the appropriate laser with the appropriate spectrometer and photomultiplier tube, but it also takes some effort to adapt the auxiliary equipment. We are told in a paper by Walrafen (from the Bell Telephone Laboratories, Murray Hill, New Jersey), when he published measurements on HDO in H_2O, partially done by Ar^+ laser excitation, partially with a mercury lamp:[111] "Laser Raman methods have not yet been refined in this laboratory such that they can replace the conventional mercury excitation methods for quantitative intensity measurements... involving changes in temperature and concentration."

Development in laser techniques proceeds rapidly, and this is warranted to provide still more powerful sources and still better adapted wavelengths for the investigation of poorly scattering samples. In such conditions, easily exchangeable equipment is of higher value than single, complete Raman spectrometers. The most significant achievements of the last years were not accomplished by these spectrometers, but by fitting a sample to a selected laser and collecting the scattered radiation into the slit of an appropriate monochromator or even spectrograph.

SAMPLE IMPROVEMENT

With laser excitation it is possible, in principle, to choose the wavelength which will produce maximum gain in scattered radiation without producing fluorescence. With double monochromators stray light is excluded, and by photoelectric recording there is the possibility of subtracting (even by use of computers) the remaining interference. It is almost unnecessary to discuss the demands raised and the practical hints given on quality and treatment of the samples in classical textbooks.[112-114] For some years to come research will probably still go on in quite a number of laboratories without all the conditions listed in the first lines of this section fulfilled. This is to be expected since only with difficult samples will the necessity of all the modern apparatus really be apparent; on the other hand results of infrared spectroscopy which are not complemented by Raman measurements will be looked on more and more in the future as being incomplete.

In the following we will give some examples from the recent literature illustrating that with poorly scattering samples care from the beginning of preparation or special treatment is important even in work aided by double monochromators or by laser excitation. The well known principles of sample handling for Raman spectroscopy are, briefly: to exclude fluorescence, any contact with rubber or grease must be excluded; to eliminate vestiges therefrom and Tyndall scattering, distillation is the only absolutely effective means. Since cracking or oxydation at higher temperature may produce fluorescent impurities, distillation *in vacuo* is often needed, even under oxygen-free nitrogen. Vacuum becomes a problem without grease, so concentrated sulfuric acid or phosphorus pentoxide liquified by hygroscopic hydration is used for wetting the ground joints at their outer rims. P_2O_5, if not resublimed, should not come into contact with samples. It will impart fluorescence also when used as a drying agent.

The glassware, and especially the sample tubes, must, of course, be clean, but only in one instance have extreme precautions been advocated for Raman spectroscopy:[115] the tube was cleaned "as it is commonly done" (which will mean a chromic acid treatment followed by splashing with doubly distilled water). Then the tube was rinsed further by condensing steam. The apparatus was constructed in such a way that whenever the steam was turned off, no air, which would contain particles of dust, could enter the tube. Drying was accomplished by a vacuum system, and the vacuum was kept to introduce the samples by evaporation through a sintered glass disk at 40° below their boiling points at atmospheric pressure. The compounds had been dried in the vapor phase by BaO. The rinsing by condensing vapors may be done easier with acetone.[116]

Special prescriptions for conditioning single compounds may be found in the earlier literature, possibly outdated because of the improved quality of commercial products. Water has always needed the most attention. At one time it was refluxed over permanganate before distillation, but only the latter is really important. Preferably, it should be done in an all-glass unit, as far as possible avoiding joints or, at least, grease, gaskets, and sealing cuffs. This apparatus should carefully be inspected to make sure that neither aerosol produced from the bursting bubbles in boiling nor creeping surface film may reach the condenser. Hornig and coworkers stated in a paper on water that they found no treatment whatsoever necessary, and that water straight from the tap was usable in recording the bands from valency and bending vibrations in studying intensity variations produced by dissolved electrolytes,[34] but this attitude has not become the generally accepted one. Water was purified and forced through a millipore filter of 0.01 μm pore size when, for the investigation of HDO dissolved in H_2O, an Ar^+ laser with the Cary model 81 spectrometer was employed.[111]

Work on Rayleigh scattering intensities, where there is no spectral discrimination of the quantity to be measured from the light scattered in other ways, requires much more effort in sample preparation, and these experiences could be of value, occasionally, to Raman spectroscopists. The all-Pyrex still for water from Fisher Catalog No. 9-107 was (in 1963) especially praised.[116,117]

At one time in such work carbon disulfide was only distilled,[118] but on one occasion[119] we had to go through the entire procedure given by Pestemer in his compilation on purification of solvents for UV spectroscopy[120] in order to get clear Raman spectra. It is always advisable to try products of other manufacturers if trouble with a particular solvent is experienced.

Hydrocarbons are percolated over silica[118] or treated as vapors[121] in order to remove traces of fluorescent impurities. But any treatment which goes beyond distillation must be looked upon as an experiment of unguaranteed success. This is especially the case with charcoal. Different batches should be tested before final application.

Compounds which cannot be distilled were often prepared—this at least seems to have become avoidable now—for Raman spectroscopy exclusively from distillable or volatile components following all the precautions concerning fluorescence and Tyndall scattering. In order to investigate the anion of tetrametaphosphimic acid,[52] we prepared tetrameric phosphornitrilic chloride, $P_4N_4Cl_8$, by separating it from $P_3N_3Cl_6$ in fractionated distillation. Then we resublimed it twice in $vacuo$ at 1 torr and recrystallized it twice from (purified) benzene. We could have transformed this starting material to the desired salt by hydrolysis with solutions of appropriate bases, but we prepared the acid, which is known to crystallize well, from the solution of the chloride in ether (purified) by shaking with water (Raman quality). Salts were prepared by adding—after other experiments—to freshly prepared solutions of the acid, the carbonates of Rb and Cs. The effort was rewarded, since from solutions with the latter cation, which could be obtained in a concentration of almost $55 g (PO_2NH)_4^-$ to $100 g H_2O$, even observation of NH vibrations, well known from infrared spectroscopy (1332 and $1305 cm^{-1}$ in-plane deformation, $742 cm^{-1}$ out-of-plane deformation), was possible with photographic exposure times of up to 50 h.

For success with these solutions filtration was also important. It seems remarkable that when using the finest sintered glass filters available, which are stated to have pore sizes from 0.7 to 1.5 μm, no improvement at all concerning residual Tyndall scattering was detected. As in other cases,[122] filtration through foils of collodium was necessary. Magat, in his investigation on water,[28] had already passed this through such membranes. The observation of metal–oxygen vibrations of aquo complexes was possible with quite unpretentious photographic work and even led to measurements of depolarization for solutions with Al^{+3}, Mg^{+2}, Zn^{+2}, and Be^{+2} ions.[123] It seems important that these were subjected to ultrafiltration, meaning filtration through collodium membranes.

In the papers on recent work by means of double monochromators or lasers there are some statements on the importance of filtration (and of previous precautions) with water when aqueous solutions of poor scattering compounds are investigated. To record bands from aluminate ions[124] in solutions of $Al(OH)_3$ in NaOH, which were $4M$ in Al^{+3}, the liquids were passed through a filter of 0.2-μm pore size immediately prior to measurement. In the first experiments with a Perkin–Elmer LR-1 Raman spectrometer high noise was experienced. This is ascribed to discrete scattering particles drifting in and out of the laser beam. After a technique for preparing clean solutions had been developed, the signal-to-noise ratio with the LR-1 was significantly improved. It is reported to have been consistently better than with a Cary model 81 instrument using mercury 4358-Å excitation, when the same spectral slit width, scanning rates, time constants, and samples were employed.

In photographic work with the Hilger E 612 apparatus on the same chemical system and on the similar one of $Zn(OH)_2$–$NaOH$[125] it was also found that blank solutions are difficult to obtain, but that it may be done by careful technique. The filter employed is the Corning 33,900 GVYKU ultrafine bacteriological filter. Distillation of water in an all-glass still was also important.

A similar investigation of alumina and silica dissolved in $NaOH$[126] relied, apparently, on centrifugation, before use, of the samples only. But from our own experience with solutions of metaphosphates,[122] it must be stated that the effects of this treatment may be only very limited generally and not comparable with improvements obtained by filtration. That there might be results is acknowledged. In the classical times of Raman spectroscopy it had already been hinted that standing for some time will allow scattering impurities to settle.[113]

Trouble with alkaline solutions is reported in older work repeatedly.[127] It is the general belief that some silicic acid is set free from glass vessels in a highly dispersed, but not dissolved, state. Walrafen took account of this when he used Teflon beakers for the preparation of germanate solutions,[128] and polyethylene bottles to store them. Glass volumetric ware was only employed for dilution, and filtration was done through sintered glass filters of very fine porosity (quality VF). It was found that minimized contact with glass reduced turbidity and gave solutions of the high optical quality required in intensity measurements. These were done with a Cary model 81 instrument.

If everything is done to avoid Tyndall scattering and fluorescing impurities, and the solutions are as concentrated as possible, and there is still too low scattered intensity, as may be the case preferably with inorganic compounds, there is still the possibility of changing to the solid state, especially to single crystals, to get information from Raman spectroscopy.

If the art of Raman distillation and filtration is going to be forgotten through the merits of laser excitation and double monochromators, there remains another challenge to laboratory skill: the art of producing large crystals. It was stated that sometimes clearer observations are possible because of band sharpening in the solid state. Furthermore, it is obvious that measurements of the ratios of depolarization, which are so easily and accurately obtained by laser excitation, require, if not solutions, then single crystals. When a laser beam is sent into a poly-crystalline solid, its polarization is lost by reflections, and possibly other effects of crystal optics, in an irregular array. Hendra[129] tried this experiment: square and octahedral halogeno complexes of Pt, Pd, and Au show one polarized line ($\rho < 0.1$) in solution, but no effect of polarization is observed when the solid salts are examined with the Cary 81 laser instrument.

Besides these general facts which require work with crystals there is a special importance in the solid state in investigations on poor scattering samples.

Theoretical reasons lead to the expectation of a higher ratio of intensities of Raman to Rayleigh scattering. The former is incoherent, but the latter is not, and so optical interferences are possible. These will produce partial extinction for Rayleigh scattering in the case of crystals, where the scattering centers are arranged in a regular pattern, contrary to the state of liquids and gases. For crystals the intensity of both the scattered radiations may be from the same order of magnitude, whereas for liquids, Raman lines generally appear at least 100 times weaker than Rayleigh lines,[130] but Raman intensities are almost the same in the liquid and in the solid state, as was shown for t-butanol.[131] (We mention this example because it is also valuable in providing the exception to the rule about band sharpening in the solid state: the hydrogen-bonded OH valency vibration could not be measured in the liquid or solid state, but only in the gas phase, where it becomes a sharp band.)

The advantage of a higher Raman-to-Rayleigh intensity ratio in the solid state is usually lost with polycrystalline samples, where the scattering of primary light by inhomogeneity exceeds by far the Rayleigh radiation. With single crystals, however, this advantage is felt very much. The author of this chapter would like to cite experiences from his laboratory[132] in this connection.

For phosphoric acid the recorded spectra (obtained with the Cary model 81 spectrometer, mercury arc excitation) were nearly equivalent to those photographed—only some minor bands were missing. A decisive improvement was possible after preparation of single crystals. The different procedures adopted for growing such crystals cannot be described here. Best results were obtained with the specimens grown right into the sample tube. Space remaining between the crystal and the walls or the front plate of the tube may be filled with an immersion liquid. We employed carbon tetrachloride. Unfortunately, crystal growing within the tube was only possible with $H_3PO_4 \cdot \frac{1}{2}H_2O$. In melts of phosphoric acids containing less water, some diphosphoric acid is always formed. This accumulated in the zone of crystal growth and stopped it. The crystal of phosphoric acid hemihydrate did not show cracks or inclusions in its usable part of 40 mm length and 8 mm diameter. It appeared homogeneous when illuminated and viewed through plates covered with polarizing foil. In the best results with polycrystalline hemihydrate only one band (at $912 \, cm^{-1}$) had been found. Exposures with the single crystal yielded 12 bands or shoulders, as they are produced by crystal field splittings. (We thank Dr. Reich from the Group for Physical Methods of Analysis at Deutsche Akademie der Wissenschaften zu Berlin, Berlin-Aldershof, for trying his best on the polycrystalline phosphoric acid with the Cary model 81 spectrometer, mercury arc excitation. Our work on the single crystal was done photographically.) Nothing was observed from hydrogen (or deuterium) valency vibrations. They are well known from infrared spectroscopy.[133] Statements on $OH \cdots O$ bending modes are not so easy, but at least no in-plane bending vibrations are observed.

A second example will show relations between single crystal and solution work which are more likely to be met generally: an investigation of potassium hydrogen maleate, $K[HOOC-CH=CH-COO]$. The Raman spectra of a cut crystal in the appropriate directions of excitation and polarization disclosed most of the intramolecular[134] and intermolecular[135] vibrations expected from the selection rules for the unit cell, and also some modes which were missing in the table of wave numbers of the ion in solution.[136] There are two vibrations of species a_1 of the free ion among those which were discovered only by the single-crystal work! The authors have not investigated or discussed whether these are OH motions.

Relevancy of optical conditions of solid samples in the case of poor scatterers may be seen well by comparing the results obtained for different modifications of ice with different appearance. Taylor and Whally[137] placed most of the samples in the tubes as small pieces, including ordinary ice, "ice I_h." Another ice, I_c, was only obtained as a powder; this one scattered too strongly for photoelectric recording, but in photographic work three bands from OH valency vibrations are revealed, as for the other modifications. Ice I_h had been investigated already by Valkov and Maslenkova.[138] They had used a larger clear rod, and had observed three lines and a broad, weak, asymmetric band in the region considered. This latter was, although mercury lines were interfering, barely seen as a suggestion in the later work. To again detect the lines which the Soviet authors had observed between 230 and 290 cm^{-1}, excitation with the mercury resonance line at 2537 Å was employed. The Canadian authors found agreement, but also stated that they could not see all the detail.

We shall close this chapter with these lines. First we had to show that further progress with poor scatterers depends on the joint application of powerful lasers and double monochromators. But we hope that we have now appeased those spectroscopists who wish to see that something remains to be done in the chemical laboratory.

REFERENCES

1. A. C. Chapman, *Spectrochim. Acta* **24A**: 1687 (1968).
2. E. Steger and R. Stahlberg, *Z. Naturforsch.* **17b**: 780 (1962); *Z. Anorg. Allgem. Chem.* **326**: 243 (1964).
3. J. C. Hisatsune, *Spectrochim. Acta* **21**: 1899 (1965).
4. H. M. Gager, J. Lewis, and M. J. Ware, *Chem. Commun.*: 116 (1966).
5. A. M. Adams, J. B. Cornell, J. L. Dawes, and R. D. W. Kemmitt, *Inorg. Nuclear Chem. Letters* **3**: 437 (1967).
6. J. P. Russel, *J. Physique* **26**: 620 (1965).
7. V. S. Ryazanov and V. S. Gorelik, *Fiz. Tverd. Tela* **10**: 1909 (1968).
8. H. Siebert, *Z. Anorg. Allgem. Chem.* **275**: 225 (1954).
9. J. Brandmüller, *Optik* **12**: 389 (1955).
10. J. R. Nielsen, *J. Opt. Soc. Am.* **37**: 494 (1947).
11. G. E. Walrafen, *J. Chem. Phys.* **40**: 3249 (1964).
12. J. R. Durig and D. W. Wertz, *J. Mol. Spectry.* **25**: 467 (1968).
13. R. W. Rudolph, R. W. Parry, and C. F. Farran, *Inorg. Chem.* **5**: 723 (1966).
14. D. A. Brown, D. Cunningham, and W. K. Glass, *J. Chem. Soc. Am.*: 1563 (1968).
15. N. W. Alcock, *J. Chem. Soc. Am.*: 1588 (1968).

16. J. A. Evans and D. A. Long, *J. Chem. Soc. Am.*: 1688 (1968).
17. J. C. Evans and G. Y.-S. Lo, *J. Phys. Chem.* **70**: 11 (1966).
18. J. Brandmüller, *Z. Angew. Phys.* **5**: 95 (1953).
19. J. C. Evans and G. Y.-S. Lo, *J. Phys. Chem.* **71**: 3942 (1967).
20. J. Shamir and H. H. Hyman, *J. Phys. Chem.* **70**: 3132 (1966).
21. J. S. Kirby-Smith and E. A. Jones, *J. Opt. Soc. Am.* **39**: 780 (1949).
22. W. E. Klee, *Z. Anorg. Allgem. Chem.* **343**: 58 (1966).
23. S. Pinchas and D. Sadeh, *J. Inorg. Nuclear Chem.* **30**: 1785 (1968).
24. G. Davidson, E. A. V. Ebsworth, G. M. Sheldrick, and L. A. Woodward, *Spectrochim. Acta* **22**: 67 (1966).
25. B. Beagly, A. G. Robiette, and G. M. Sheldrick, *Chem. Comm.*: 601 (1967).
26. H. Siebert, J. Eints, and E. Fluck, *Z. Naturforsch.* **236**: 1006 (1968).
27. G. Bolla, *Nuovo Cimento (N.S.)* **9**: 290 (1932); **10**: 101 (1933); **12**: 243 (1935); *J. Chem. Phys.* **6**: 225 (1938).
28. M. Magat, *J. Phys. Radium* (7) **5**: 347 (1934).
29. M. Magat, *Ann. Phys. (Paris)* **6**: 109 (1936).
30. J. H. Hibben, *J. Chem. Phys.* **5**: 994 (1937).
31. P. K. Narayanaswamy, *Proc. Indian Acad. Sci.* **27A**: 311 (1948).
32. W. R. Busing and D. F. Hornig, *J. Phys. Chem.* **65**: 284 (1961).
33. J. W. Schultz and D. F. Hornig, *J. Phys. Chem.* **65**: 2131 (1961).
34. T. T. Wall and D. F. Hornig, *J. Chem. Phys.* **43**: 2079 (1965).
35. G. R. Harrison, R. C. Lord, and J. R. Loofbourow, *Practical Spectroscopy*, Prentice-Hall, New York (1948 and later 1954), p. 63.
36. G. W. Walrafen, *J. Chem. Phys.* **36**: 1035 (1962).
37. D. E. Irish and T. F. Young, *J. Chem. Phys.* **43**: 1765 (1965).
38. T. F. Young and G. E. Walrafen, *Trans. Faraday Soc.* **57**: 34 (1961).
39. N. G. Zarakhani, M. I. Vinnik, *Zh. Fiz. Khim.* **37**: 2550 (1963).
40. J. Brandmüller and H. Moser, *Einführung in die Ramanspektroskopie*, translated to Russian by G. V. Perechubova and Kh. E. Stepina, M. M. Sushchinski, ed. Moscow, MIR (1964).
41. R. F. Stamm, C. F. Salzmann, and T. Mariner, *J. Opt. Soc. Am.* **43**: 119 (1953).
42. E. Ziegler and A. G. Hoffmann, paper presented to Raman Symp., Magdeburg 1966; *Österr. Chem. Ztg.* **68**: 319 (1967).
43. J. Brandmüller and H. Moser, *Einführung in die Ramanspektroskopie*, Dr. Dietrich Steinkopff Verlag, Darmstadt (1962).
44. H. Gerding and H. Houtgraaf, *Rec. Trav. Chim.* **72**: 21 (1953).
45. K. Balsubrahmanyam and D. Nanis, *J. Chem. Phys.* **42**: 676 (1965).
46. J. S. Ziomek and M. D. Zeidler, *J. Mol. Spectry.* **11**: 163 (1963).
47. J. R. Durig, S. F. Bush, and E. E. Mercker, *J. Chem. Phys.* **44**: 4238 (1966).
48. J. Goubeau, *Z. Physik. Chem.* **B45**: 237 (1940).
49. Yu. I. Kotov and V. M. Tatevski, *Opt. i Spektroskopiya* **14**: 443 (1963).
50. J. R. Durig and J. W. Clark, *J. Chem. Phys.* **48**: 3216 (1968).
51. J. S. Kittelberger and D. F. Hornig, *J. Chem. Phys.* **46**: 3099 (1967).
52. K. Lunkwitz, Thesis, Technische Universität Dresden, 1964; cf. E. Steger and K. Lunkwitz, *J. Mol. Structure*, in press.
53. H. W. Schrötter, *Z. Angew. Phys.* **12**: 275 (1960); see also, Brandmüller and Moser, reference 43, p. 157.
54. G. Heintz and A. Simon, *Z. Physik Chem.* **216**: 67 (1961).
55. A. Simon, H. Kriegsmann, and E. Steger, *Z. Physik. Chem.* **205**: 190 (1956).
56. W. Otting, *Der Ramaneffekt und seine analytische Anwendung*, Springer, Berlin (1952).
57. A. Simon, *Kolloid-Z.* **85**: 8 (1938).
58. H. A. Szymanski, R. Yelin, and L. Marabella, *J. Chem. Phys.* **47**: 1877 (1967).

59. J. R. Durig, D. W. Wertz, B. R. Mitchell, F. Block, and J. M. Greene, *J. Phys. Chem.* **71**: 3815 (1967).
60. K. Hirano, *Bull. Chem. Soc. Japan* **41**: 731 (1968).
61. R. C. Lord and G. J. Thomas, *Biochim. Biophys. Acta* **142**: 1 (1967); *Spectrochim. Acta* **23A**: 2551 (1967).
62. W. Sawodny, K. Niedenzu, and J. W. Dawson, *J. Chem. Phys.* **45**: 3155 (1966).
63. B. Gilbert, *Bull. Soc. Chim. Belges* **76**: 493 (1967).
64. R. S. Krishnan Vidya, *J. Gujerat Univ.* **6**: 158 (1963); C. A. **64** 10613 f.
65. F. Rasetti, *Z. Physik* **61**: 598 (1930).
66. L. Cecchi, *J. Physique* **26**: 469 (1965).
67. S. E. Ostroy and E. A. McGinnis, *Proc. Pennsylvania Acad. Sci.* **35**: 172 (1961).
68. H. Wolff and D. Staschewski, *Z. Elektroch. Ber. Bunsenges. Phys. Chem.* **65**: 840 (1961).
69. H. Wolff and H. Ludwig, *Z. Elektrochim. Ber. Bunsenges. Phys. Chem.* **70**: 474 (1966).
70. J. H. R. Clarke and C. Solomon, *J. Chem. Phys.* **47**: 1823 (1967).
71. L. A. Woodward and M. J. Taylor, *J. Chem. Soc.*: 407 (1962).
72. J. H. Carpenter, W. J. Jones, R. W. Jotham, and L. H. Long, *Chem. Comm.*: 881 (1968).
73. G. G. Dumas, *Compt. Rend.* **B266**: 1589 (1968).
74. J. P. Marsault and G. G. Dumas, *Compt. Rend.* **B264**: 782 (1967).
75. R. J. Obremsky, C. W. Brown, and E. R. Lipincott, *J. Chem. Phys.* **49**: 185 (1968).
76. M. Ito, *Spectrochim. Acta* **21**: 2063 (1965).
77. E. V. Perezhina and Zh. Zh. Raskin, *Dokl. Akad. Nauk SSSR* **150**: 1022 (1963).
78. G. Karagounis and R. Issa, *Nature* **195**: 1196 (1962).
79. P. J. Hendra and E. J. Loader, *Nature* **216**: 789 (1967).
80. P. J. Hendra and E. J. Loader, *Nature* **217**: 637 (1968).
81. R. P. Young and N. Sheppard, *J. Catalysis* **7**: 223 (1967); *Trans. Faraday Soc.* **63**: 2291 (1967).
82. G. Zerbi and P. J. Hendra, *J. Mol. Spectry.* **27**: 17 (1968).
83. P. J. Hendra and H. A. Willis, *Chem. Commun.*: 225 (1968).
84. F. P. Linn and J. L. König, *J. Mol. Spectry.* **9**: 228 (1962).
85. S. Krimm and C. G. Opsaker, *Spectrochim. Acta* **21**: 1165 (1965).
86. M. Tosumi, T. Shimanouchi, and T. Miyazawa, *J. Mol. Spectry.* **9**: 261 (1962).
87. J. M. Schachtschneider and R. G. Snyder, *Spectrochim. Acta* **19**: 117 (1963).
88. J.-P. Mon, *Compt. Rend.* **B266**: 244 (1968).
89. T. C. Damen, S. P. S. Porto, and B. Tell, *Phys. Rev.* **142**: 570 (1966).
90. J. P. Kaminow, R. C. C. Leite, and S. P. S. Porto, *J. Phys. Chem.* **26**: 2085 (1965).
91. S. P. S. Porto and R. S. Krishan, *J. Chem. Phys.* **47**: 1009 (1967).
92. R. S. Krishnan, *Proc. Indian Acad. Sci.* **26A** · 450 (1947).
93. C. H. Perry, J. H. Fertel, and T. F. McNelly, *J. Chem. Phys.* **47**: 1619 (1967).
94. W. G. Nilson and J. G. Skinner, *J. Chem. Phys.* **48**: 2240 (1968).
95. J.-P. Mon, *J. Physique* **26**: 611 (1965).
96. J.-P. Mon, *Compt. Rend.* **B266**: 1222 (1968).
97. J. P. Russel, *Proc. Phys. Soc.* (*London*) **85**: 194 (1965).
98. M. Krauzman, *Compt. Rend.* **B266**: 1224 (1968).
99. R. S. Krishnan, *Proc. Indian Acad. Sci.* **A26**: 450 (1947).
100. G. B. B. M. Sutherland, *Proc. Roy. Soc.* (*London*) **A141**: 535 (1933).
101. A. I. Stekhanov and M. B. Eliashberg, *Opt. i Spektroskopiya* **10**: 174 (1961).
102. Nguyen Xuan Xinh, A. H. Maradudin, and R. A. Coldwell-Horsfall, *J. Physique* **26**: 717 (1965).
103. R. Kaiser and P. Möckel, *Phys. Letters* **25A**: 749 (1967).
104. J. P. Hurrell, S. P. S. Porto, T. C. Damen, and S. Mascaenhas, *Phys. Letters* **26A**: 194 (1968).

105. W. Holzer, W. F. Murphy, H. J. Bernstein, and I. Rolfe, *J. Mol. Spectry.* **26**: 543 (1968).
106. H. Stammreich, D. Bassi, O. Sala, and H. Siebert, *Spectrochim. Acta* **13**: 192 (1958).
107. L. D. Vinh, J. Reynond, and R. Lafont, *Compt. Rend.* **B263**: 192 (1966).
108. M. S. Mathur, C. A. Frenzel, and E. B. Bradley, *J. Mol. Structure* **2**: 429 (1968).
109. D. A. Brown, D. Cunningham, and W. K. Glass, *Spectrochim. Acta* **24A**: 965 (1968).
110. N. Krauzman, *Compt. Rend.* **B263**: 922 (1966).
111. G. E. Walrafen, *J. Chem. Phys.* **48**: 244 (1968).
112. K. W. F. Kohlrausch, *Der Smekal-Raman-Effekt,* Springer, Berlin (1931), Ergänzungsband 1931–1937, Berlin (1938).
113. K. W. F. Kohlrausch, "Raman-Spektren," in: A. Eucken und K. L. Wolf, *Hand und Jahrbuch der chemischen Physik,* Akad. Verlags-Ges., Leipzig (1943), Vol. 9, Sect. 6.
114. J. H. Hibben, *The Raman Effect and its Chemical Applications,* Reinhold, New York (1947).
115. H. D. Mallory, *J. Chem. Phys.* **18**: 898 (1950).
116. J. P. Kratochvil, M. Kerker, and L. E. Oppenheimer, *J. Phys. Chem.* **67**: 1097 (1963).
117. J. P. Kratochvil, M. Kerker, and L. E. Oppenheimer, *J. Chem. Phys.* **43**: 914 (1963).
118. D.-J. Coumen, E. L. Mackor, and J. Hijmans, *Trans. Faraday Soc.* **60**: 1539 (1964).
119. R. Stahlberg, Diploma work, Dresden (1962).
120. M. Pestemer, *Angew. Chem.* **63**: 118 (1951); **67**: 746 (1955).
121. M. R. Fenske, W. G. Braun, R. V. Wiegand, D. Quiggle, L. H. McCormick and D. H. Rank, *Anal. Chem.* **19**: 700 (1947).
122. A. Simon and E. Streger, *Z. Anorg. Allgem. Chem.* **277**: 209 (1954); E. Steger and A. Simon, *Z. Anorg. Allgem. Chem.* **291**: 76 (1957).
123. A. de Silveira, M. A. Marques, and N. M. Marques, *J. Mol. Spectry.* **20**: 88 (1966).
124. L. A. Carreira, V. A. Maroni, J. W. Swaine, Jr., and R. C. Plumb, *J. Chem. Phys.* **45**: 2216 (1966).
125. E. R. Lippincott, J. A. Psellos, and M. C. Tobin, *J. Chem. Phys.* **20**: 536 (1952).
126. J. Turkevich and S. Ciborowski, *J. Phys. Chem.* **71**: 3208 (1967).
127. A. Simon and G. Schulze, *Z. Anorg. Allgem. Chem.* **242**: 313 (1939).
128. G. E. Walrafen, *J. Chem. Phys.* **42**: 485 (1965).
129. P. J. Hendra, *Spectrochim. Acta* **23A**: 2871 (1967).
130. F. Matossi, *Der Ramen-Effekt,* Friedr. Vieweg u. Sohn, Braunschweig (1959).
131. F. Perzl and H. Moser, *J. Mol. Spectry.* **26**: 237 (1968).
132. K. Herzog, Thesis, Technische Universität Dresden (1966).
133. K. Herzog and E. Steger, *J. Inorg. Nuclear Chem.* **27**: 1429 (1965).
134. C. Bardot, G. Marignan, and J. Maillols, *Compt. Rend.* **B265**: 53 (1967).
135. C. Bardot, G. Marignan, and J. Maillols, *Compt. Rend.* **B266**: 611 (1968).
136. C. Bardot, G. Marignan, and J. Maillols, *Compt. Rend.* **C264**: 825 (1967).
137. M. J. Taylor and E. Whally, *J. Chem. Phys.* **40**: 1660 (1964).
138. V. I. Valkov and G. L. Maslenkova, *Opt. i Spektroskopiya* **1**: 881 (1956).

Appendix

Comments on the Derivation of the Dispersion Equation for Molecules*

Gene P. Barnett† and A. C. Albrecht

Chemistry Department
Cornell University
Ithaca, New York

INTRODUCTION

It seems useful to present here a derivation of the dispersion equation, and from it an expression for the intensity of scattered light, for molecules. The equation has been derived for atoms in standard texts,[1,2] obtaining the Kramers–Heisenberg equation for the photon–atom collision cross section. It has also been derived with some approximations for molecules both quantum mechanically, via a more general formalism than the one to be developed here,[3] and semiclassically.[4] We will present the derivation with a minimum of mathematical complexity and within the framework of nonrelativistic quantum theory. In the process we will try to expose as clearly as possible the approximations that are generally needed and used in practice.

First we define the system in terms of a zeroth-order Hamiltonian, presenting the standard quantum mechanical description of the radiation field, and then develop in detail the operator describing the interaction of a molecule with an external electromagnetic field. The dispersion equation and the intensity formula for molecular scattering of an incident radiation source are then developed. Finally, the molecular polarizability is identified by analogy to classical theory.

THE ZEROTH-ORDER PROBLEM

The system is defined as light of photon frequency $\omega_0 = 2\pi\nu_0$ incident on an isolated, freely rotating, molecule. The molecule undergoes a transition from the molecular eigenstate Ψ_n of energy E_n

*This work has been assisted by a grant from the National Science Foundation.
†Present address: Chemistry Department, Denver University, Denver, Colorado.

to eigenstate Ψ_m of energy E_m, and the incident light is scattered to frequency ω_f (for elastic scattering $m = n$ and $\omega_f = \omega_0$). The energy conservation relationship for this process is

$$E_m - E_n = \hbar(\omega_0 - \omega_f) \tag{1}$$

For a molecule exposed to an external radiation field, one can write a nonrelativistic Hamiltonian

$$H = H^0 + H' \tag{2}$$

where the zeroth-order Hamiltonian H^0 describes an unperturbed system of molecule plus radiation field,

$$H^0 = H_{mol} + H_{rad} \tag{3}$$

and H' describes the interaction of the two.

The molecular Hamiltonian is written in the usual form

$$H_{mol} = T_E + T_N + V \tag{4}$$

where T_E and T_N represent the electronic and nuclear kinetic energies, respectively, and V denotes the potential energy for electron–electron, electron–nuclear, and nuclear–nuclear interactions. The space of the Hamiltonian is described by the coordinates $\mathbf{r}$ and $\mathbf{Q}$, where $\mathbf{r} = \mathbf{r}_1$, $\mathbf{r}_2, \ldots, \mathbf{r}_N$ is a shorthand notation for the coordinate vectors of the N electrons relative to a molecule-fixed coordinate system and $\mathbf{Q} = \mathbf{Q}_1, \ldots, \mathbf{Q}_M$ represents the normal coordinates for the M nuclei, excluding translation and rotation. For the state n, the Schrödinger equation is

$$H_{mol}\Psi_n(\mathbf{r}, \mathbf{Q}) = E_n\Psi_n(\mathbf{r}, \mathbf{Q}) \tag{5}$$

The molecular rotation is treated, later in the development, by averaging over all orientations.

QUANTUM DESCRIPTION OF THE RADIATION FIELD

The electromagnetic field is considered as an ensemble of non-interacting harmonic oscillator modes;[5] the μth mode containing n_μ photons, all having energy $\hbar\omega_\mu$, polarization $\hat{\mathbf{e}}_\mu$, and propagation $\pm \mathbf{k}_\mu$. The propagation vector has magnitude

$$|\mathbf{k}_\mu| = \omega_\mu/c \tag{6}$$

where c is the velocity of light, and it is perpendicular to the polarization vector

$$\hat{e}_\mu \cdot \mathbf{k}_\mu = 0 \tag{7}$$

The Hamiltonian for the ensemble is

$$H_{rad} = \sum_\mu H_\mu \tag{8}$$

where

$$H_\mu = 2\omega_\mu^2 q_\mu q_\mu^* \tag{9}$$

and q, q^* are the time-dependent, complex oscillator amplitudes. The eigenfunction for the unperturbed field is written in the product form

$$\Psi_{rad}(q_1, q_2, \ldots) = u_{n_1}(q_1) u_{n_2}(q_2) \ldots \tag{10}$$

where there are n_1 photons of frequency ω_1, polarization $\hat{e}_1$, and propagation $\pm\mathbf{k}_1$, etc. The individual oscillator functions satisfy

$$H_\mu u_{n_\mu}(q_\mu) = n_\mu \hbar \omega_\mu u_{n_\mu}(q_\mu) \tag{11}$$

and the total field energy is then

$$E_{rad} = \sum_\mu n_\mu \hbar \omega_\mu \tag{12}$$

where the sum is over all modes in the ensemble. The field eigenfunction should be symmetric to exchange of any two coordinates q_μ, q_λ since photons obey Bose–Einstein statistics. However, as we will consider explicitly only two oscillator modes, this complication need not concern us here. The functions $u_n(q)$ are the well known harmonic oscillator eigenfunctions having the properties

$$\langle u_n | u_m \rangle = \delta_{nm} \tag{13}$$

$$q u_n = [n\hbar/2\omega]^{\frac{1}{2}} u_{n-1} \tag{14}$$

$$q^* u_n = [(n+1)\hbar/2\omega]^{\frac{1}{2}} u_{n+1} \tag{15}$$

where q and q^* may be thought of as operators that describe photon annihilation and creation, respectively.

The properties of the electromagnetic field are described in terms of a vector potential $\mathbf{A}$ that satisfies the equations

$$\nabla^2 \mathbf{A} - 1/c^2 \frac{\partial^2}{\partial t^2} \mathbf{A} = 0 \tag{16}$$

$$\operatorname{div} \mathbf{A} = 0 \tag{17}$$

With the normalization condition

$$\langle A_\lambda | A_\mu \rangle = 4\pi c^2 \delta_{\lambda\mu} \tag{18}$$

we have

$$A_\lambda = [4\pi c^2]^{\frac{1}{2}} \hat{e}_\lambda \, e^{i k_\lambda \cdot r} \tag{19}$$

and

$$A = \sum_\lambda \{ q_\lambda A_\lambda + q_\lambda^* A^* \} \tag{20}$$

where the sum is over all modes in the field. For the case of interest to us, frequencies ω_0 and ω_f, we can write

$$A = (4\pi c^2)^{\frac{1}{2}} [\hat{e}_0 (q_0 e^{i k_0 \cdot r} + q_0^* e^{-i k_0 \cdot r}) + \hat{e}_f (q_f e^{i k_f \cdot r} + q_f^* e^{-i k_f \cdot r})] \tag{21}$$

From equation (18) and the boundary conditions imposed on A and its derivatives, periodicity in a finite volume V (and let $V = 1$), one finds that the number of photon states available to a photon with propagation vector $\pm k_\lambda$ is

$$\frac{k_\lambda^2 \, dk_\lambda \, d\Omega_\lambda}{(2\pi)^3} = \frac{\omega_\lambda^2 \, d\omega_\lambda \, d\Omega_\lambda}{(2\pi c)^3} \tag{22}$$

The density of photon states ρ_λ per unit energy $d(\hbar\omega_\lambda)$ in the interval $\hbar\omega_\lambda$ to $\hbar\omega_\lambda + d(\hbar\omega_\lambda)$ is, then,

$$\rho_\lambda = \frac{\omega_\lambda^2 \, d\Omega_\lambda}{(2\pi c)^3 \hbar} \tag{23}$$

THE INTERACTION TERM

The interaction between the molecule and the external field is due to the coupling of the vector potential A with the moving charged particles—electrons of charge $-e$ and mass μ_e, nuclei of charge eZ_a and mass μ_a. Thus, the momentum operators in the kinetic energy terms of equation (4), which are

$$T_E = \frac{1}{2\mu_e} \sum_{i=1}^{N} p_i^2 \tag{24}$$

and

$$T_N = \sum_{a=1}^{M} \frac{1}{2\mu_a} P_a^2 \tag{25}$$

transform as

$$\mathbf{p}_i \rightarrow \left[\mathbf{p}_i - \frac{e}{c} \mathbf{A}(\mathbf{r}_i) \right] \tag{26}$$

and

$$\mathbf{P}_a \rightarrow \left[\mathbf{P}_a + \frac{eZ_a}{c} \mathbf{A}(\mathbf{Q}_a) \right] \tag{27}$$

Accordingly, the additional terms in the Hamiltonian which describe the interaction

$$H' = H'_E + H'_N \tag{28}$$

are the electronic term

$$H'_E = \sum_{i=1}^{N} \left[\frac{-e}{\mu_e c} \mathbf{p}_i \cdot \mathbf{A}(\mathbf{r}_i) + \frac{e^2}{\mu_e c^2} \mathbf{A}^2(\mathbf{r}_i) \right] \tag{29}$$

and the nuclear term

$$H'_N = \sum_{a=1}^{M} \left[\frac{eZ_a}{\mu_a c} \mathbf{P}_a \cdot \mathbf{A}(\mathbf{Q}_a) + \frac{e^2 Z_a^2}{\mu_a c^2} \mathbf{A}^2(\mathbf{Q}_a) \right] \tag{30}$$

In practice the interaction operator is usually simplified by making the following approximations: the nuclear term is generally neglected altogether, and the electric dipole approximation is made. The electric dipole approximation requires that $|\mathbf{k}| |\mathbf{r}|$ be much less than unity so the terms $e^{\pm i\mathbf{k} \cdot \mathbf{r}}$ in equation (21) can be approximated by the first term of the series

$$e^{\pm i\mathbf{k} \cdot \mathbf{r}} = 1 + (\pm i\mathbf{k} \cdot \mathbf{r}) + \tfrac{1}{2}(\pm i\mathbf{k} \cdot \mathbf{r})^2 + \cdots \tag{31}$$

In this approximation the $\mathbf{A}^2$ terms given no contribution to Raman scattering due to orthogonality within the set of zeroth-order molecular eigenstates. The implications of neglecting the nuclear term H'_N will be discussed in the following section.

THE DISPERSION EQUATION

The transition probability for light scattering requires matrix elements for the perturbation or interaction operator, equation (28), in terms of the unperturbed states which are the eigenstates of H_0 described earlier. The original state is

$$\Psi_0 = \Psi_n(\mathbf{r}, \mathbf{Q})u_{n_0}(q_0)u_{n_f}(q_f) \tag{32}$$

the final state

$$\Psi_F = \Psi_m(\mathbf{r}, \mathbf{Q})u_{n_0-1}(q_0)u_{n_f+1}(q_f) \tag{33}$$

and their energies are

$$E_0 = E_n + n_0\hbar\omega_0 + n_f\hbar\omega_f \tag{34}$$

$$E_F = E_m + (n_0 - 1)\hbar\omega_0 + (n_f + 1)\hbar\omega_f \tag{35}$$

Actually, we could set $n_f = 0$ since we are considering incident radiation of frequency ω_0 only (and we are neglecting black-body radiation effects); it is instructive, however, to carry it further in the discussion. The transition probability through second order in perturbation theory is given by

$$W_{FO} = \frac{2\pi}{\hbar}\rho_f \left| H'_{FO} + \sum_{I,II} \left\{ \frac{H'_{FI}H'_{IO}}{E_0 - E_I} + \frac{H'_{FII}H'_{IIO}}{E_0 - E_{II}} \right\} \right|^2 \tag{36}$$

where H'_{FO} is the first-order term, the second-order term is summed over two types of intermediate or virtual states, and ρ_f is the density of final states for scattered photons. To first order, the scattering can be described as the simultaneous absorption and emission of photons of frequency ω_0 and ω_f, respectively. The second-order description of scattering involves two types of intermediate states: I) the photon ω_0 is absorbed and the molecule forms an intermediate state, say Ψ_r; the photon ω_f is emitted (spontaneously) and the final state Ψ_m is formed; II) conversely, the molecule may spontaneously emit the photon ω_f and form a different intermediate state $\Psi_{r'}$, then the incident photon is absorbed to form the final state. The matrix elements are

$$H'_{FO} = \langle \Psi_m u_{n_0-1}u_{n_f+1}|H'|\Psi_n u_{n_0}u_{n_f}\rangle \tag{37}$$

$$H'_{FI} = \langle \Psi_m u_{n_0-1}u_{n_f+1}|H'|\Psi_r u_{n_0-1}u_{n_f}\rangle \tag{38}$$

$$H'_{IO} = \langle \Psi_r u_{n_0-1}u_{n_f}|H'|\Psi_n u_{n_0}u_{n_f}\rangle \tag{39}$$

$$H'_{FII} = \langle \Psi_m u_{n_0-1}u_{n_f+1}|H'|\Psi_r u_{n_0}u_{n_f+1}\rangle \tag{40}$$

$$H'_{IIO} = \langle \Psi_r u_{n_0}u_{n_f+1}|H'|\Psi_n u_{n_0}u_{n_f}\rangle \tag{41}$$

A graphical description of the two-photon process in terms of Feynman diagrams has been discussed by Peticolas.[7]

At this point two approximations are generally introduced. The first is the electric dipole approximation; the consequence of this is that the first-order matrix element and the A^2 terms in the second-order matrix elements are zero for Raman scattering as mentioned earlier. The second approximation is to neglect completely the nuclear term H'_N in the perturbation operator, equation (28). When one does this, it turns out that in the expression for the transition probability, equation (36), two types of matrix element products are being ignored. They are nuclear–nuclear and electron–nuclear terms, respectively, of the type

$$\langle \Psi_m u_{n_0-1} u_{n_f+1} | H'_N | \Psi_r u_{n_0-1} u_{n_f} \times \Psi_r u_{n_0-1} u_{n_f} | H'_N | \Psi u_{n_0} u_{n_f} \rangle \quad (42)$$

$$\langle \Psi_m u_{n_0-1} u_{n_f+1} | H'_E | \Psi_r u_{n_0-1} u_{n_f} \times \Psi_r u_{n_0-1} u_{n_f} | H'_N | \Psi_n u_{n_0} u_{n_f} \rangle \quad (43)$$

A brief inspection of these terms within the Born–Oppenheimer approximation and using harmonic oscillator nuclear wave functions shows that equation (42) is nonzero only for Raleigh scattering and for overtone and combination contributions to Raman scattering. Equation (43) is likely to be nonzero for Rayleigh and overtone or combination Raman scattering (i.e., vibrational Raman) and it is also nonzero for electronic Raman scattering through only one intermediate state. The relative contributions of these terms, equations (42) and (43), as compared to electron–electron type terms for vibrational Raman scattering in the ground electronic state, has been analyzed in terms of powers of the electron–nuclear mass ratio by Tang and Albrecht.[8] It seems safe to conclude that in the general case the nuclear term H'_N probably contributes little to the scattering process.

Now that these two approximations have been discussed, and since they do not in general seriously restrict the theory we will for convenience adopt them here and continue with the derivation.

Performing the integrations over the q variables of radiation space first, using the relationships of equations (13)–(15), the matrix elements become

$$H'_{FI} = \left[\frac{(n_f+1)\hbar}{2\omega_f} \frac{4\pi e^2}{\mu_e^2} \right]^{\frac{1}{2}} \langle \Psi_m | -\sum_{i=1}^{N} \mathbf{p}_i \cdot \hat{\mathbf{e}}_f | \Psi_r \rangle \quad (44)$$

$$H'_{10} = \left[\frac{n_0 \hbar}{2\omega_0} \frac{4\pi e^2}{\mu_e^2} \right]^{\frac{1}{2}} \langle \Psi_r | -\sum_{i=1}^{N} \mathbf{p}_i \cdot \hat{\mathbf{e}}_0 | \Psi_n \rangle \quad (45)$$

$$H'_{FII} = \left[\frac{n_0 \hbar}{2\omega_0} \frac{4\pi e^2}{\mu_e^2} \right]^{\frac{1}{2}} \langle \Psi_m | -\sum_{i=1}^{N} \mathbf{p}_i \cdot \hat{\mathbf{e}}_0 | \Psi_r \rangle \quad (46)$$

$$H'_{110} = \left[\frac{(n_f + 1)\hbar}{2\omega_f} \frac{4\pi e^2}{\mu_e^2} \right]^{\frac{1}{2}} \langle \Psi_r | - \sum_{i=1}^{N} \mathbf{p}_i \cdot \hat{\mathbf{e}}_f | \Psi_n \rangle \qquad (47)$$

Suppressing the sum over i and introducing the notation

$$\mathbf{p}_{rn} = \langle \Psi_r(\mathbf{r}, \mathbf{Q}) | \mathbf{p} | \Psi_n(\mathbf{r}, \mathbf{Q}) \rangle \qquad (48)$$

the transition probability, equation (36), is approximated as

$$W_{FO} = \frac{2\pi}{\hbar} \rho_f \left[\frac{(n_f + 1)\hbar}{2\omega_f} \frac{n_0 \hbar}{2\omega_1} \right]^2 \frac{\pi^2 e^4}{\mu_e^4}$$

$$\times \left| \sum_r \left\{ \frac{\mathbf{p}_{mr} \cdot \hat{\mathbf{e}}_f \, \mathbf{p}_{rn} \cdot \hat{\mathbf{e}}_0}{E_n - E_r + \hbar\omega_0} + \frac{\mathbf{p}_{mr} \cdot \hat{\mathbf{e}}_0 \, \mathbf{p}_{rn} \cdot \hat{\mathbf{e}}_f}{E_m - E_r - \hbar\omega_0} \right\} \right|^2 d\Omega_f \qquad (49)$$

Introducing the expression for ρ_f, and setting $n_f = 0$, we have the transition probability for the Raman process with the incident photon beam being inelastically scattered into the solid angle $d\Omega_f$ as

$$W_{FO} \, d\Omega_f = \frac{\omega_f}{\omega_0} \frac{n_0 e^4}{c^3 \mu_e^4} \left| \sum_r \left\{ \frac{\mathbf{p}_{mr} \cdot \hat{\mathbf{e}}_f \, \mathbf{p}_{rn} \cdot \hat{\mathbf{e}}_0}{E_n - E_r + \hbar\omega_0} + \frac{\mathbf{p}_{mr} \cdot \hat{\mathbf{e}}_0 \, \mathbf{p}_{rn} \cdot \hat{\mathbf{e}}_f}{E_m - E_r - \hbar\omega_0} \right\} \right|^2 d\Omega_f$$

$$(50)$$

If we had not set $n_f = 0$ we could instead consider incident radiation at both frequencies ω_0 and ω_f, and would be able to observe both spontaneous and induced scattered radiation at frequency ω_f. The familiar Kramers–Heisenberg dispersion formula is obtained from equation (50) if we set $n_0 = 1$ and consider one photon incident on one molecule, divide by the relative velocity or incident flux (approximately c), and obtain the collision cross section

$$d\phi = \frac{\omega_f}{\omega_0} \frac{e^4}{c^4 \mu_e^4} \left| \sum_r \left\{ \frac{\mathbf{p}_{mr} \cdot \hat{\mathbf{e}}_f \, \mathbf{p}_{rn} \cdot \hat{\mathbf{e}}_0}{E_n - E_r + \hbar\omega_0} + \frac{\mathbf{p}_{mr} \cdot \hat{\mathbf{e}}_0 \, \mathbf{p}_{rn} \cdot \hat{\mathbf{e}}_f}{E_m - E_r - \hbar\omega_0} \right\} \right|^2 \qquad (51)$$

We prefer instead to consider an incident beam of radiation with an average intensity I_0, in units of energy per unit area and per unit time, given by

$$I_0(\hbar\omega) \, d(\hbar\omega_0) = \bar{n}_0 \rho_0 c \hbar\omega_0 \, d(\hbar\omega_0) \qquad (52)$$

$I_0(\hbar\omega)$ gives the incident intensity distribution in the interval $\hbar\omega_0$ to $(\hbar\omega_0 + d\hbar\omega_0)$ and I_0 is the integral of $I_0(\hbar\omega)$ over $d(\hbar\omega_0)$. As opposed to our idealized zeroth-order description of n_0 photons per oscillator of energy $\hbar\omega_0$ we have to consider an average number $\bar{n}_0$ of photons per oscillator with energy in $d(\hbar\omega_0)$. As a consequence, the number of photons in the initial state is $\bar{n}_0 \rho_0 \, d(\hbar\omega_0)$ which we substitute for n_0 in equation (50). This allows us to express equation (50) in terms of the incident intensity I_0 via equation (52). Now, multiplying both sides of equation (50) by $\hbar\omega_f$ to express the results in terms of scattered intensity, we get the result

$$I_f \, d\Omega_f = \frac{\omega_f^2}{\omega_0^2}\left(\frac{e}{\mu_e c}\right)^4 I_0 \left|\sum_r \left\{\frac{\mathbf{p}_{mr} \cdot \hat{\mathbf{e}}_f \, \mathbf{p}_{rn} \cdot \hat{\mathbf{e}}_0}{E_n - E_r + \hbar\omega_0} + \frac{\mathbf{p}_{mr} \cdot \hat{\mathbf{e}}_0 \, \mathbf{p}_{rn} \cdot \hat{\mathbf{e}}_f}{E_m - E_r - \hbar\omega_0}\right\}\right|^2 d\Omega_f$$

(53)

The radiation in the energy interval $d(\hbar\omega_f)$ scattered into the solid angle $d\Omega_f$ is given as the intensity I_f, in units of energy per second.

We still have to account for the rotational motion of the molecule. Expanding the dot products give terms $\cos^2 \theta_0$ and $\cos^2 \theta_f$. The angle θ_0 is between the dipole momentum vector $\mathbf{p}$ and the incident polarization vector $\hat{\mathbf{e}}_0$. By performing the average

$$\frac{1}{4\pi} \int \cos^2 \theta_0 \, d\Omega_0 = \tfrac{1}{3}$$

(54)

we treat molecular rotation by averaging over all orientations of $\mathbf{p}$. The solid angle $d\Omega_f$ can be transformed from its reference position about the scattered polarization vector to an angle with respect to the scattered propagation vector. We are free to consider $\mathbf{p}$ in the plane formed by the vectors $\mathbf{k}_f$ and $\hat{\mathbf{e}}_f$; thus,

$$\cos \theta_f = \sin\left(\frac{\pi}{2} - \theta_f\right) = \sin \theta_f$$

(55)

since $\mathbf{k}_f \cdot \hat{\mathbf{e}}_f = 0$. Now our expression in equation (53) becomes

$$I_f \, d\Omega_f = \frac{\omega_f^2}{\omega_0^2} \frac{e^4}{3c^4 \mu_e^2} I_0 \left|\sum_r \left\{\mathbf{p}_{mr} \, \mathbf{p}_{rn}[(E_n - E_r + \hbar\omega_0)^{-1}\right.\right.$$

$$\left.\left. + (E_m - E_r - \hbar\omega_0)^{-1}]\right\}\right|^2 \sin^2 \theta_f \, d\Omega_f$$

(56)

This gives the intensity scattered into the solid angle $d\Omega_f$ about the scattered propagation vector $\mathbf{k}_f$. The total scattered intensity in the energy interval $d(\hbar\omega_f)$ due to incident radiation in the energy interval $d(\hbar\omega_0)$ is, then,

$$I_f = \frac{\omega_f^2}{\omega_0^2} \frac{e^4 2^3 \pi}{3^2 c^4 \mu_e^4} I_0 \left|\sum_r \left\{\mathbf{p}_{mr} \, \mathbf{p}_{rn}[(E_n - E_r + \hbar\omega_0)^{-1}\right.\right.$$

$$\left.\left. + (E_m - E_r - \hbar\omega_0)^{-1}]\right\}\right|^2$$

(57)

THE MOLECULAR POLARIZABILITY

When a molecule is placed in an electric field it forms an induced dipole moment in addition to any permanent dipole moment that may exist. The magnitude of the induced moment is proportional to the field strength for weak fields and is given as

$$\mu_{ind} = \alpha \cdot \mathbf{E}_0 \tag{58}$$

where α is called the molecular polarizability. Classically, the induced dipole oscillates at the same frequency as the field and will absorb incident light and reemit it with no shift in frequency, thus giving rise to Rayleigh scattering. The polarizability changes slightly with the vibrational and rotational motion of the molecule and, as a consequence, can also emit light at frequencies shifted from the incident frequency. This gives rise to the Stokes and anti-Stokes lines of Raman scattering.[9] The scattered intensity in all cases due to a freely rotating induced dipole is

$$I = \frac{\omega^4}{3^2 c^3} |\mu_{ind}|^2 \tag{59}$$

Using the standard relation

$$I_0 = \frac{c}{8\pi} |\mathbf{E}_0|^2 \tag{60}$$

we can rewrite equation (57) as

$$I_f = \frac{\omega_f^4}{3^2 c^3} \left| \frac{e^2/\mu_e^2}{\omega_0 \omega_f} \sum_r \left\{ \mathbf{p}_{mr} \, \mathbf{p}_{rn} [(E_n - E_r + \hbar\omega_0)^{-1} \right. \right.$$
$$\left. \left. + (E_m - E_r - \hbar\omega_0)^{-1}] \right\} \mathbf{E}_0 \right|^2 \tag{61}$$

and thus identify the molecular polarizability as

$$\alpha_{mn} = \frac{1}{\omega_0 \omega_f} \frac{e^2}{\mu_e^2} \sum_r \left\{ \mathbf{p}_{mr} \, \mathbf{p}_{rn} [(E_n - E_r + \hbar\omega_0)^{-1} \right.$$
$$\left. + (E_m - E_r - \hbar\omega_0)^{-1}] \right\} \tag{62}$$

The diagonal term α_{nn} is the polarizability tensor for Rayleigh scattering and the nondiagonal terms $\alpha_{mn}, m \neq n$, are the Raman polarizability scattering tensors.

It is often more convenient, but not necessarily more accurate,[10] to work with α_{mn} in the dipole length form of the matrix elements, e.g., $\mathbf{r}_{mr}$, rather than the dipole momentum form used here. The transformation, outlined by Dirac,[1] is left as an informative exercise for the reader. The result is

$$\alpha_{mn} = \sum_r \{\mathbf{r}_{mr}\,\mathbf{r}_{rn}[(E_n - E_r + \hbar\omega_0)^{-1} + (E_m - E_r - \hbar\omega_0)^{-1}]\} \quad (63)$$

REMARKS

Both the polarizability and the scattered intensity could instead have been left in terms of the polarization vectors for incident and scattered light. Then one can directly study properties of various components of the scattering tensor α_{mn} in terms of selection rules and symmetry properties.

The theory presented here does not account for radiative damping phenomena.[11] The more general quantum mechanical description introduces the damping constant and natural line breadth directly in the theory. In our equations incident radiation at a resonance frequency gives an infinite scattering intensity; we can avoid this difficulty by adding to each denominator the term $i\gamma_r$, where γ_r is the natural breadth of the molecular state Ψ_r.

Recent reviews on various experimental and theoretical applications of Raman scattering are available.[7,12] Additional references on the quantum theory of radiation which are especially of pedagogical value are Armstrong[13] and Fowler.[14]

REFERENCES

1. P. A. M. Dirac, *The Principles of Quantum Mechanics*, 4th ed., Oxford University Press, London (1958), Chapt. 10.
2. W. Heitler, *The Quantum Theory of Radiation*, 3rd ed., Oxford University Press, London (1954), Chapt. 5.
3. See Ref. 1, Chapt. 12; Ref. 2, Chapts. 4 and 6; H. F. Hameka, *Advanced Quantum Chemistry*, Addison-Wesley, Reading, Mass. (1965), Chapts. 11 and 12; G. Placzek, in: E. Marx, *Handbuch der Radiologie*, Akademische Verlagsgessellschaft, Leipzig (1939), Vol. 2, Part 2, p. 209, translated by A. Werbin, 1959, Office of Technical Services, Department of Commerce, Washington 25, D.C.
4. J. Behringer and J. Brandmüller, *Z. Elektrochem.* **60**: 643 (1956).
5. See Ref. 2, Chapt. 2, and Ref. 1.
6. J. C. Slater, *Quantum Theory of Atomic Structure*, McGraw-Hill, New York (1960), Vol. 1, Chapt. 6.
7. W. L. Peticolas, in: H. Eyring, *Annual Review of Physical Chemistry*, Annual Reviews, Palo Alto (1967), Vol. 18, p. 233.

8. J. Tang and A. C. Albrecht, in: H. A. Szymanski, *Raman Spectroscopy*, Vol. 2, Plenum Press, New York (1970), Chapt. 2.

9. G. Herzberg, *Molecular Spectra and Molecular Structure*, Van Nostrand, Princeton (1950), Vol. 1, p. 82.

10. H. A. Bethe and E. E. Salpeter, *Quantum Mechanics of One- and Two-Electron Atoms*, Springer-Verlag, Berlin (1957), Chapt. 4a.

11. See Ref. 2, p. 181.

12. H. A. Szymanski, *Raman Spectroscopy*, Plenum Press, New York (1967).

13. B. H. Armstrong, in: J. L. Magee and H. Aroeste, *Thermal Radiation Phenomena*, Plenum Press, New York (1969), Vol. 2, Chapt. 3.

14. G. N. Fowler, in: D. R. Bates, *Quantum Theory*, Academic Press, New York (1962), Vol. 3, Chapt. 2.

Index